Operator Education and Regulation; Safe Mobility for Older Persons; Traffic Law Enforcement, and Occupant Protection 2018

AA005722

Transportation Research Record: Journal of the Transportation Research Board

Volume 2672, Issue 33

ISBN: 978-1-5108-7835-8

Printed from e-media with permission by:

Curran Associates, Inc.
57 Morehouse Lane
Red Hook, NY 12571

Some format issues inherent in the e-media version may also appear in this print version.

Published by Sage Publications USA

ISBN (Print) 978-1-5108-7835-8
ISBN 2018 Printed Set (All Issues) 978-1-5108-7735-1

Printed by Curran Associates, Inc. (2019)

Additional copies of this publication are available from:

Curran Associates, Inc.
57 Morehouse Lane
Red Hook, NY 12571 USA
Phone: 845-758-0400
Fax: 845-758-2633
Email: curran@proceedings.com
Web: www.proceedings.com

Volume 2672 Number 33 December 2018

TRR
JOURNAL OF THE TRANSPORTATION RESEARCH BOARD

Contents

Operator Education and Regulation; Safe Mobility for Older Persons; Traffic Law Enforcement; and Occupant Protection 2018

Articles

---

* Membership as of August 2018.

TRB

TRANSPORTATION RESEARCH BOARD

CORE SPONSORS

Transportation Departments of the 50 States and the District of Columbia

Federal Government

U.S. Department of Transportation
- Federal Aviation Administration (FAA)
- Federal Highway Administration
- Federal Motor Carrier Safety Administration (FMCSA)
- Federal Railroad Administration (FRA)
- Federal Transit Administration
- National Highway Traffic Safety Administration (NHTSA)
- Office of the Assistant Secretary of Transportation for Research and Technology (OST-R)

U.S. Air Force Civil Engineer Center (AFCEC)
U.S. Coast Guard (USCG)
U.S. Department of Energy
U.S. Department of the Interior (DOI)
U.S. Environmental Protection Agency (EPA)

Nongovernmental Organizations

American Public Transportation Association (APTA)
Association of American Railroads (AAR)
California Air Resources Board

The National Academies of
SCIENCES · ENGINEERING · MEDICINE

The **National Academy of Sciences** was established in 1863 by an Act of Congress, signed by President Lincoln, as a private, nongovernmental institution to advise the nation on issues related to science and technology. Members are elected by their peers for outstanding contributions to research. Dr. Marcia McNutt is president.

The **National Academy of Engineering** was established in 1964 under the charter of the National Academy of Sciences to bring the practices of engineering to advising the nation. Members are elected by their peers for extraordinary contributions to engineering. Dr. C. D. Mote, Jr., is president.

The **National Academy of Medicine** (formerly the Institute of Medicine) was established in 1970 under the charter of the National Academy of Sciences to advise the nation on medical and health issues. Members are elected by their peers for distinguished contributions to medicine and health. Dr. Victor J. Dzau is president.

The three Academies work together as the National Academies of Sciences, Engineering, and Medicine to provide independent, objective analysis and advice to the nation and conduct other activities to solve complex problems and inform public policy decisions. The Academies also encourage education and research, recognize outstanding contributions to knowledge, and increase public understanding in matters of science, engineering, and medicine.

Learn more about the National Academies of Sciences, Engineering, and Medicine at **www.national-academies.org.**

The **Transportation Research Board** is one of seven major programs of the National Academies of Sciences, Engineering, and Medicine. The mission of the Transportation Research Board is to increase the benefits that transportation contributes to society by providing leadership in transportation innovation and progress through research and information exchange, conducted within a setting that is objective, interdisciplinary, and multimodal. The Board's varied activities annually engage about 7,000 engineers, scientists, and other transportation researchers and practitioners from the public and private sectors and academia, all of whom contribute their expertise in the public interest. The program is supported by state transportation departments, federal agencies including the component administrations of the U.S. Department of Transportation, and other organizations and individuals interested in the development of transportation.

Learn more about the Transportation Research Board at **www.TRB.org**.

Article

TRR
JOURNAL OF THE TRANSPORTATION RESEARCH BOARD

Transportation Research Record
2018, Vol. 2672(33) 1–10

DOI: 10.1177/0361198118759950
journals.sagepub.com/home/trr
SAGE

Situation Awareness Fast-Tracking, Including Identifying Escape Routes (SAFER): Evaluation of the Impact of SAFER on Learner Driver Situation Awareness Skills

Bridie Scott-Parker[1,2,3], Leigh Wilks[1,2,3], and Bonnie Huang[1,2,3]

Abstract

Despite a plethora of education, engineering, and enforcement-related intervention, the pernicious problem that is young driver road safety remains of global interest. Compared with more experienced drivers, young novice drivers have been found to have deficits in situation awareness skills (SAS), which is an essential repertoire of knowledge and abilities in perceiving, comprehending, and appropriately responding to a breadth of driving risks (projection). Current practice requirements in Queensland, Australia, do not incorporate SAS-specific training for parents, the most common supervisor of novice drivers. This study evaluates the impact of SAFER, a SAS-acquisition acceleration "game" in which parents foster SAS in their child during the period *before* licensure, on novice driver SAS *at* learner licensure. Sixty parent–pre-learner dyads were recruited from the Sunshine Coast and randomly allocated to intervention (n = 30) and control (n = 29). Using a SAS-based coding taxonomy, SAS was measured via simulator-based verbal commentary protocol at learner licensure as part of a larger longitudinal project. Intervention learners exhibited significantly greater SAS (perception/comprehension/projection of breadth of driving risks), than control learners. Intervention learners exhibited significantly less perception, and considerably greater perception/comprehension/projection SAS than intervention parents. Currently, in Queensland's licensing program there is limited support for parents/other supervisors of learner drivers, and no SAS-focused intervention is available. SAFER is an innovative SAS-acquisition acceleration intervention that has been shown to build SAS even before the young novice is licensed to drive. A larger state-wide pilot is in development to explore the merit of incorporating SAFER within Queensland's graduated driver licensing program.

The intractable global problem that is young driver road safety is evidenced as a persistent overrepresentation in road crashes despite a plethora of targeted and more general engineering, education, and enforcement interventions (*1*). In Australia in 2016, youth aged 17–25 years had a fatality rate of 9.0/100,000 population, compared to 5.5/100,000 population of road users aged 40–64 years, with young males fatally injured (13.4) at almost three times the rate of young females (4.4) (*2*). Moreover, young driver crashes involve more than the young driver; for example, in Queensland in 2016, young drivers were involved in 33.8% of hospitalization crashes and 24.7% of hospitalization crashes across the Australian state (*3*), despite comprising 12.7% of the licensed population (*4*).

To mitigate road safety risks, jurisdictions such as Queensland have implemented a multistage graduated driver licensing (GDL) program. The first – learner – license phase is entered after passing a learner theory test, which can be taken, at the earliest, at the age of 16 years. The learner phase is characterized by mandatory practice requirements including a minimum of 100 certified logbook hours (a minimum of which 10 hours must be at night) over a minimum 12-month period (*5*). Perhaps unsurprisingly give the cost of professional driving lessons, parents are the most common driving supervisor (*6, 7*). Perplexingly, however, parents are typically "left to their own devices," with no option for gaining training in how and what to teach their young novice

[1]Adolescent Risk Research Unit (ARRU), Sunshine Coast Mind and Neuroscience – Thompson Institute, University of the Sunshine Coast, Birtinya, Queensland, Australia
[2]School of Social Sciences, Faculty of Arts, Business, and Law, University of the Sunshine Coast, Sippy Downs, Queensland, Australia
[3]Sustainability Research Centre, Faculty of Arts, Business, and Law, University of the Sunshine Coast, Sippy Downs, Queensland, Australia

Corresponding Author:
Address correspondence to Bridie Scott-Parker: bscottpa@usc.edu.au

driver, and a dearth of resources beyond compliance with the road rules (e.g., compliance with regulatory signs such as posted speed limits), safe vehicle operation (e.g., sharing the road with vulnerable road users such as cyclists and pedestrians), and vehicle safety (pre-driving checks of indicators and windscreen wipers) (*8, 9*).

Parents are also largely unaware that, as experienced drivers, they have developed a wealth of situation awareness skills (SAS) which are critical – and likely automatic (*10*) – for road safety. While the precise attributes and/or mechanisms of cognitive processes pertaining to the concepts of *hazard detection, hazard perception*, and thus *situation awareness* can differ according to the manner in which SAS is examined, there is some consensus in the literature that SAS mean that the experienced driver can (*a*) correctly perceive and identify hazards in the driving environment (*perception*), (*b*) understand why the hazards are a road safety-related problem (*comprehension*), and (*c*) anticipate how these hazards may impact on their road safety (*projection*), thus informing the regulation of their behavior to ameliorate or mitigate these hazards altogether (*11, 12*). Unsurprisingly, young and novice drivers have SAS deficits (*13, 14*); therefore, it seems logical that GDL and other licensing programs that require mandatory practice before unsupervised on-road driving incorporate a training program for the main driving supervisor – parents, in the case of Queensland, Australia – so that learner drivers receive optimal training to accelerate the acquisition of SAS.

It is noteworthy that in jurisdictions such as Queensland, there is limited *pre*-license (and thus, pre-on-road driving) systematic training for young and novice drivers. Young and novice drivers may have exposure to programs such as the RYDA (Rotary Youth Driver Awareness) program, in which key road safety-related factors such as traveling speed, the influence of personality and social factors, and hazardous behavior such as driving distracted are discussed during a whole-of-class visit to a central location for a fee. The RYDA is not available in all areas of Queensland, and SAS is not a key education component. It is also noteworthy that in jurisdictions such as Queensland, there is no *pre*-license (and thus, pre-on-road driving) systematic training for parents of soon-to-be young novice drivers. Experiencing this situation for herself and her own teen, and given her expertise in young and novice driver road safety, the first author piloted an SAS-acquisition acceleration program – *situation awareness fast-tracking, including identifying escape routes* (SAFER) – with her pre-license teen, with superior SAS objectively noted by a professional driving instructor just a handful of hours into learner license practice. Given the success of the $n = 1$ pilot, the first author secured government funding for a larger pilot study.

For the larger pilot study, SAFER training emphasizes the breadth of risks related to – and thus the realms within which SAS require development – the key factors of other road users (e.g., to the side of trucks), infrastructure (e.g., avoid telegraph posts), maneuvers (e.g., drive left in an emergency), and exposure (e.g., if route unclear because of sun glare, do not plan to drive in that direction). The importance and identification of potential escape routes was also demonstrated. Briefly, the SAFER game follows the following basic procedure (refer also to Figure 1):

1. Before commencing the SAFER game:
   a. The parent demonstrates what potential and actual driving hazards look like (*perception*), why they are a hazard (*comprehension*), and how to regulate their own behavior to eliminate or ameliorate the hazard (*projection*).
   b. The parent and the teen agree on a prize if the teen wins by the following:
      i. They are able to exhaustively demonstrate SAS through perception, comprehension, and projection of risks *to themselves* (i.e., to all occupants of the vehicle in which they are traveling); AND
      ii. They are able to exhaustively demonstrate SAS through perception, comprehension, and projection of risks *to other road users* (i.e., to all road users within a reasonable distance who could be harmed by the vehicle in which they are traveling).
2. The parent asks the teen to start identifying hazards and to use the *perception, comprehension, and projection* (PCP) approach. For example, using the "road users" image in Figure 1:
   a. Teen: "There is a car coming toward us in the oncoming lane, and I am watching his lane position in case he is distracted or avoids a risk I can't yet see, such as a kangaroo; if I need to avoid a crash with this driver I can drive a little to the left but I am blocked in a smooth exit as there is a gutter, so I will travel in the furthest left of the lane to minimize the risk of collision. There is a car in front of me slowing for the roundabout so I am slowing down also. There is a road worker standing at the entrance to the roundabout – he has his back to me and may not realize I am even behind him so I am really slowing down so I can safely brake in an emergency in case he steps backward or to his left without checking; there are road work vehicles at the left of the roundabout – they may be parked there and so the lane may be shut, but I am going to drive through the roundabout very slowly while keeping an eye on them in case they drive forward as part of their work and do not check for vehicles already on the roundabout."
   b. Parent: "Great job, but it looks like there is another car behind the front car in our oncoming lane, so I will also watch that car and see how

Figure 1. SAFER game training resources pertaining to road users (top), infrastructure (middle), and exposure (bottom).

that driver is behaving; this will influence my driving and where we can drive in case of an emergency also; if that second car stays on the roundabout, I will need to give way so I am slowing down until I can see in which direction that driver is traveling. Also, the road work truck on the left looks like it is parked in the median strip, but by doing this he is blocking a clear view of the drivers in the lane to his left; so I am going to drive slowly and have a look past this road work truck as I get closer, just in case there is a car that does not see our vehicle and I need to brake in an emergency."

c. In this instance, the parent is the winner of the game, given they could identify other driving-related risks, explain these risk to their teen, and explain the potential impact of these risks on their driving and escape behavior.

Figure 2. An example of day-time driving footage in an urban area.

The development of SAFER, the baseline SAS evaluation of the $n = 60$ parent–young driver dyads ($n = 30$ intervention; $n = 60$ young drivers and $n = 59$ parents, as two parents had participating triplet young drivers), and the process evaluation of the SAFER intervention by parents has been reported elsewhere (*15*). Of note, at baseline parents were found to exhibit considerably greater comprehension and projection SAS compared to pre-license drivers, with no significant difference in SAS (perception, comprehension, and projection) between parents in the intervention and control groups, and between learners in the intervention and control groups. For the process evaluation, there was strong support for SAFER, with nearly every one of the 30 intervention parents agreeing that helping their child to develop SAS was beneficial for their road safety, that they were prepared to play SAFER with their child, and that they know how to improve their own SAS also (*15*). The night-time SAS and the escape-route skills of the participants is currently under analysis and is not considered in the article. This manuscript explores the day-time SAS of young driver participants now the majority have progressed to their learner license and thus the intervention learners have been playing the SAFER game with their parents for a period of approximately 6 months.

Method

Participants at Learner Licensure

As at June 3, 2017, 46 of the 60 participants ($n = 31$ male young drivers; $n = 23$ male parents) had completed the second stage of the study (teens progressed from pre-license drivers to learner license). While most pre-learner drivers were accompanied by their parent, one was accompanied by his grandfather and another was accompanied by his sister. Removal of these (age) outliers revealed no significant differences in age between parents in the control and intervention conditions (M (SD) = 46.9 (5.3), range 35–57; M (SD) = 46.3 (5.6), 27–58; respectively). Parents had an average of 29.3 (SD = 4.2) years' driving experience (intervention: 28.7 (3.6); control: 29.9 (4.7)).

Simulator Video Presentation and Procedure

Three GoPro Hero 3+ Black cameras were mounted on a vehicle internally to capture real-world driving footage of a 15-minute day-time driving segment (Figure 2). Clips were trimmed to fit the dimensions of the Immerse cave simulation studio, a three-wall projection room with total projection area of 19.4 m long by 2.25 m high (10,548 × 1200 pixels) (see also *15–17*). The study was advertised through presentations and follow-up emails at local high schools, in a local newspaper article, and via a local radio discussion concerning young driver road safety more generally, and sought to recruit pre-license teens and their intended main driving supervisor. Parent–teen dyads attended the university together. Participation in the baseline and subsequent measures required approximately 1.25 hours for each dyad, while participation in the SAFER training required approximately 50–60 minutes for intervention group parents. After consenting to participate and completing a baseline survey exploring demographics and driving experience, the dyads, individually as parent participant and teen participant, were shown an example of verbal commentary and given the opportunity to practice before proceeding to the Immerse studio for the simulated driving activity. Verbal commentary protocol requires participants to "think aloud" as if they were the driver in the video, verbalizing where they were looking and what they were paying attention to as they watched the driving video. In this way, we were able to elucidate where the participants' attention was focused and/or drawn throughout the driving task, thereby gaining an understanding of their situation awareness within the various driving contexts. For the practice, three television screens were assembled at 90° to each other, driven by a commodity computer, and participants viewed a training clip.

SAFER Intervention Training

The following summarizes the group intervention activity for intervention group parents.

1. Provide an overview of crash risks for young novice drivers.
2. Discussion around the importance and nature of SAS, and that parents' SAS differs considerably from their young novice teen's SAS; escape-route identification skills (ERIS), and that the intervention will teach them how to develop SAS and ERIS in their teen from pre-license through the learner phase.
3. Exemplars of SAS (and ERIS) within the realms of other road users (e.g., pedestrians [to the side of trucks]), infrastructure (e.g., roundabouts [avoid telegraph

posts]), maneuvers (e.g., merging [drive left in an emergency]), and exposure (e.g., dusk [if you cannot see because of glare, drive in a different direction]) (see Figure 1 for example visual stimulus).

4. A demonstration of SAFER using previously unseen driving footage.
5. An opportunity to practice with the presenter and fellow group participants with previously unseen driving footage.
6. An opportunity to ask questions/provide comments (*11*).
7. Note that parents in the control group did not undergo any group-based training or receive any information on the importance of SAS and ERIS.

Coding SAS

Nearly 24 hours of day-time verbal commentary at the point of learner licensure was recorded and transcribed. Given the highly time-consuming task that is manually coding SAS (two coders, third person as tie-breaker – for example, intervention learners 75.9% inter-rater reliability), for the purposes of the current manuscript, SAS was analyzed for a specific section of the real-world day-time driving footage. A 76.7-second high-risk driving segment was selected, featuring the car merging as it descended a rise from a low-speed local road (60 km/h), joining a flat local council motorway as an additional lane (rapidly accelerating to 100 km/h), before merging into one lane immediately before merging into two lanes of a state highway, a road that had just reduced its posted speed limit from 110 km/h to 100 km/h. Thus, this segment exposed participants to complex infrastructure (dis/appearing lanes, trees obstructing views, traveling down/uphill), a variety of vehicles (sedans, utilities, SUVs, trucks), exposure (sunny main high/motorway, trees shading approach to second high-speed merge), and maneuvers (merging).

All commentaries were transcribed verbatim. The entire commentaries for this segment were coded according to whether SAS were evidenced as:

(a) *perception* only (P; e.g., green signs, 100 zone);
(b) *perception* and *comprehension* (PC; e.g., big sign about entering the motorway; we're coming up to the corner so now I have to indicate); or
(c) *perception, comprehension*, and *projection* (PCP; e.g., We can turn at any time but we need to take care that nothing is coming on my right; a red light, says stop, I need to stop here).

Sentences in which the entire content was related to the perception, perception/comprehension, or perception/comprehension/projection of infrastructure, vehicles, exposure, and/or maneuvers were coded in their entirety. For example: "Truck in front of me is indicating that he's going to move into the right-hand lane so why is he doing that? It probably means that there's someone slower in front of him so I'll need to be aware that I will be behind a slow vehicle or more than one as it happens."

In contrast, sentences in which the participant moves between levels of situation awareness are coded separately. For example: "This is a school zone, construction work on the right-hand side, there could be heavy vehicles, and so forth" is split so that "This is a school zone" is coded as *P*, while "construction work on the right-hand side, there could be heavy vehicles, and so forth" is coded as *PC*. Verbal transcripts were analyzed using the Leximancer™ version 4 qualitative content analysis software. Based on the use of text representations of natural language, which identifies themes, concepts, and relationships between them, Leximancer™ uses program algorithms to build a thesaurus of concepts according to text characteristics of word counts, proximity, and salience (*18*). The identification of shared and unique concepts thus provides further insight into SAS.

Results

SAS of Control Learners Compared to Intervention Learners

Independent *t*-tests (reporting results pertaining to equal variance not assumed for a significant Levene's test) revealed that there was a significant difference in the average number of words pertaining to PCP for control learners compared to intervention learners ($t(42) = -2.49, p < 0.05$), with a considerably greater number of words uttered by the intervention learners. Table 1 summarizes the average number of words uttered in each SAS category, and Figure 3 shows the proportions of P/PC/PCP words used by intervention and control learners, with findings demonstrating that the intervention learners have substantially greater SAS than the control learners. An analysis of concepts in Leximancer revealed that seven concepts were shared between control and intervention learners, with four concepts (two each) differing between the groups (Table 2). The shared concepts again relate to important general road features; however, *highway* is a shared concept among the learners and *coming* is a unique concept for the control learners. For control learners, *car* was used in relation to other cars (e.g., *so can see those cars coming up on my right here*) and *lane* was used in relation to location (e.g., *end of my lane so I would have to merge to the right*). For intervention learners, *lane* was the primary concept (e.g., *merge into the other lane and make sure there's no one behind so that we can do it safely*), clearly exhibiting a substantial difference in SAS in the two learner groups.

SAS of Intervention Learners Compared to Intervention Parents

There was a significant difference in the average number of words pertaining to P for intervention learners compared to

Table 1. Average Number of Words According to Perception, Perception/Comprehension, and Perception/Comprehension/Projection

		M	SD	*p* value
Intervention Learners compared to Control Learners				
Perception	Control	28.05	17.490	0.537
	Intervention	23.84	27.260	
Perception/Comprehension	Control	49.26	25.90	0.062
	Intervention	34.44	24.44	
Perception/Comprehension/Projection	Control	39.00	36.33	0.017
	Intervention	80.28	64.96	
Intervention Learners compared to Intervention Parents				
Perception	Learners	23.84	27.26	0.022
	Parents	48.16	43.67	
Perception/Comprehension	Learners	34.44	24.44	0.091
	Parents	48.32	31.83	
Perception/Comprehension/Projection	Learners	80.28	64.96	0.176
	Parents	59.68	36.97	

Note: M = mean; SD = standard deviation.

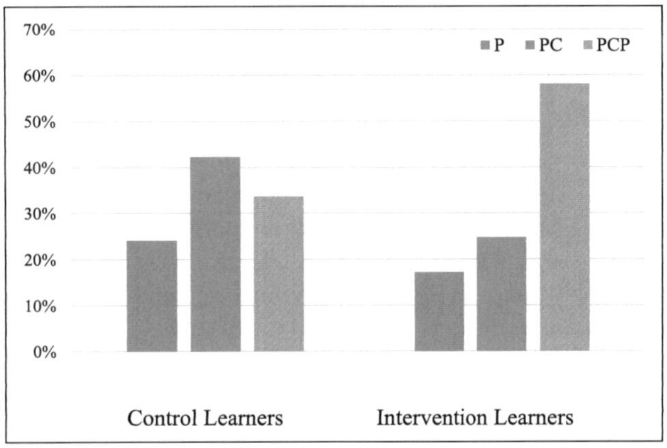

Figure 3. Proportion of perception, perception/comprehension, and perception/comprehension/projection words used by control learners and intervention learners.

intervention parents ($t(48) = -2.36$, $p < 0.05$), with parents uttering more words on average. There was significant difference in the average number of words pertaining to PC and PCP. Table 1 summarizes the average number of words uttered in each SAS category, and Figure 4 shows the proportions of P/PC/PCP words used by intervention learners and parents. Thus it appears that learners have much greater situation awareness – as evidenced by the greater proportion of PCP – than their parent driving supervisors, notwithstanding that statistical significance was not reached. An analysis of concepts in Leximancer revealed that six concepts were shared between intervention learners and parents, and that six concepts differed, with three unique concepts each (Table 2). The shared concepts again relate to important, general road features (such as *lane* and *car*) and safety-critical maneuvers such as *merge/merging* and *traveling speed*. The unique concepts for intervention parents relate to the general road environment, including *traffic* that is *coming* and in *front* of the vehicle, while pleasingly the intervention learners unique concepts contains a key situation awareness construct – *sure* – which suggests they are surveying and checking their environment more than once for (potentially) hazardous road users, maneuvers, and exposure.

General Discussion

The wicked problem that is global overrepresentation of young drivers in road crashes has generated a wealth of engineering, education, and enforcement-focused interventions. GDL is an intervention that arguably meets all three criteria in the Queensland, Australia context: Parents are the most common educator of the young and novice driver (*6, 7*), parents act as enforcers of GDL and general road rules in their supervisory role during the learner license phase (*19*), and as enforcers of GDL and road rules through the P1 phase such as by monitoring compliance with passenger limits (*19–22*); and the P1 phase of the Queensland GDL has specific vehicle engine power limits to dissuade young and novice drivers from engine-related risks such as reckless driving and other risky driving behaviors (*4*). Similarly, the critical role parents play in GDL programs is apparent in jurisdictions beyond Queensland, Australia (*23*).

Furthermore, it appears that there is a missed opportunity within the Queensland GDL to provide parents with engaging, informative, and easily applied evidence-based support to develop essential driving skills and abilities not only during the learner license phase, but during the period immediately preceding on-road driving – the pre-license phase. Interventions that specifically target parents have been found to be effective in improving the driving behaviors of young novice drivers (e.g., Checkpoints [*24*], TeenDrivingPlan [*22, 25*], Steering Teens Safe [*26*]); interventions that are designed to build hazard perception and other SAS in young novice drivers have also had some success (*27–30*). The results of the learner

Table 2. Frequency of Concepts for Learners and Parents

Control Group Learners and Intervention Group Learners				Intervention Group Learners and Intervention Group Parents			
Shared concepts				Shared concepts			
Shared concepts	Frequency control learners	Frequency intervention learners		Shared concepts	Frequency intervention learners	Frequency intervention parents	
Lane	41	58		Lane	58	74	
Car	51	42		Car	42	86	
Left	16	29		Left	29	43	
Merging	20	28		Merging	28	29	
Merge	35	28		Merge	28	40	
Speed	24	27		Speed	27	31	
Highway	15	18					
Unique concepts				Unique Concepts			
Control learners	Frequency	Intervention learners	Frequency	Intervention learners	Frequency	Intervention parents	Frequency
Coming	18	Sure	34	Sure	34	Coming	42
Motorway	17	Road	20	Road	20	Traffic	37
				Highway	18	Front	33

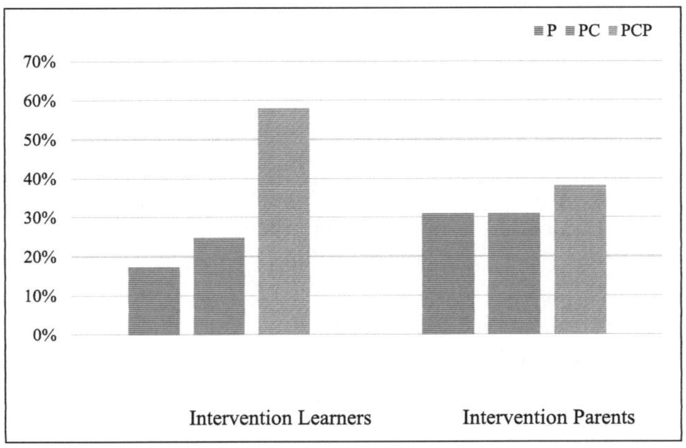

Figure 4. Proportion of perception, perception/comprehension, and perception/comprehension/projection words used by intervention learners and intervention parents.

evaluation of SAFER further support the case for targeting parents and targeting SAS, with additional evidence for the intervention efficacy in targeting the pre-license period.

Overwhelmingly, the pre-driving SAFER intervention has equipped the intervention learners with far deeper, broader, and thus safer, SAS than the control learners (note that there were no significant differences in the proportion of PCP for control learners and parents), despite the study's small sample size. These differences critically are around PCP, suggesting that the intervention learners can scan their environment and identify the various hazards (noting that in Queensland's GDL there is an online hazard perception test that must be passed before they progress from the P1 to the P2 license, (*5*)), they understand why the road user, infrastructure, or exposure-related variables are a hazard, and they can regulate their interaction with that variable to minimize that hazard. It is noteworthy that the intervention learners actually built these skills *before* they drove behind the wheel of a car, with SAS measured at the moment they progressed to a learner license. In many respects this is groundbreaking, particularly within the context that Queensland's GDL does not even expect, nor measure, the most basic SAS skill – perception – until they are trying to enter the P2 license phase.

Interestingly, both intervention and control learners do not seem to have been prompted to gain their learner license earlier than expected as a consequence of participating in this longitudinal project. Rather, the opposite effect may have happened. All teen participants were recruited at approximately 15.5 years of age, with a learner license able to be gained at 16 years. At the time of these analyses, which were completed more than 6 months post-recruitment of the last participant, 14 participants were yet to progress to their learner license. The road safety benefit of delaying licensure is well-recognized as one of the road safety benefits of GDL programs (*31*), and may be a road safety benefit of SAFER.

Strengths and Limitations

The SAFER pilot project has several strengths, including a process evaluation of parent and teen perspectives on the appeal and usability of SAFER throughout pre-license and learner license training and supervision (*15*), and a sequential, longitudinal impact evaluation of SAS 6 months pre-licensure and throughout key driving milestones, including at learner licensure, after 25 hours of learner driving (which includes GoPro-captured driving lesson content), and at P1 licensure (*32, 33*). SAS was captured via an established simulator-based verbal commentary technique (*34, 35*), with the same visual stimulus presented at each interval. Transcription and coding revealed that all commentary in the selected clip pertained to the driving context. While parents generally may have a more elaborate vocabulary in relation to perception, comprehension, and projection of driving risks, it is likely that intervention group learners may develop a similar lexicon as a consequence not only of playing SAFER during the pre-license period, but as a consequence of being taught to drive within the paradigm of SAFER's principles of risk and risk mitigation, as a learner driver. The participants were recruited from the Sunshine Coast locale, and that the day-time stimulus was captured in the Sunshine Coast means it is likely participants are familiar with the road environs. Given the extended intervals between data capture (a minimum of 6 months), and that participants are either actively playing SAFER (during the pre-license phase) or actively learning to drive with parents using a SAFER-styled approach to teaching (learner license phase), it is unlikely that participants will experience practice effects and thus inflated SAS as a result of seeing the same footage up to three times during the life of the study. The low attrition from the longitudinal project – almost 9 months in, just two participants appear to have withdrawn (one implicitly) from the project – is testament to the strong parental engagement in the project. This may reflect the biased nature of sampling, however, as parents self-selected for themselves and by extension their teenager, to participate in the longitudinal project. At this time it is unclear how frequently parents played SAFER with the pre-license teen; thus conclusions about the optimal (and minimum) practice requirements cannot be drawn at this time.

Future Research Directions

The next measurement interval for SAFER participants (both control and intervention groups) involves the recording of one week's in-car driving practice at approximately 25 hours of logged driving. The teaching content and style will be manually coded, and the techniques and lesson content of parents in the intervention group will be compared to the control group. This evaluation will provide insight into how, and if, SAFER is operationalized during the learning-to-drive phase. SAS data pertaining to the night-time driving footage have been captured, and coding is currently underway for a comparable highly complex driving segment. The SAS of parents and learners in both the control and intervention groups will be compared in the same way for the day-time SAS, as reported in this manuscript. In addition, an SAS categorization neural network is currently in development so that the SAS for learners and parents during the entire night-time and day-time driving clips can be elucidated. A coding taxonomy for ERIS (including content and related infrastructure, road users, and exposure) is currently in development based on night-time and day-time driving clips, and is particularly relevant for young driver road safety given their overrepresentation in tailgating crashes (*36*). SAFER is currently being revised and applied to suit a larger Queensland population (*37*) and other high-risk driving populations, including learner drivers with an Austism Spectrum Disorder (*38*) and older drivers (*39*).

Concluding Remarks

The wicked problem of young driver road crashes suggests that innovative intervention is required. Targeting parents, as an indirect way to target teens, is one such innovative approach. Throughout the learner license phase, parents are "left to their own devices" in many respects, with limited guidance about what and how to teach their child. Moreover, typically parents do not begin to consider learning to drive and teaching to drive until their child has already gained their learner license. SAFER harnesses the potential of parents as future providers of instruction and supervision during the final 6 months leading up to learner license acquisition, a time during which SAFER can become a habitual driving activity, and the dialog can begin to become prefaced with "You will soon be in this driver's seat." The situation awareness of intervention learners in a highly complex driving circumstance is clearly broader and deeper than the SAS of control learners, suggesting that these can be acquired pre-licensure. Moreover, the substantial PCP skills apparent in intervention learners relate to a breadth of risks associate with infrastructure, other road users, and driving exposure – all critical risk factors for and contributors to young novice driver crashes. While several analyses are pending (e.g., an evaluation of the SAFER intervention impact on SAS during night-time driving and on escape-route identification during day-time and night-time driving), it appears that SAFER is an effective SAS-acquisition acceleration program, particularly during the pre-license period.

Acknowledgments

Special thanks to research assistant Ms. Jeanne Dayton for assisting with developing and implementing SAFER training and evaluations; Mr. Nick Hansen for contributions toward simulation resources and situation awareness coding; Ms. Wanda Griffin, Mr. Tyrone Huckstepp, Ms. Kate English for contributing to the literature review, data analysis, and reporting; interns Ms. Jo Goodlet and Ms. Emily Johnson for assisting with transcribing and coding verbal commentaries, and project manager Dr. Thea Vanags for her management of the SAFER project and team. The first author is supported by a National Health and Medical Research Council Research Fellowship. The SAFER pilot is funded by the Queensland Motor Accident Commission (MAIC).

References

1. World Health Organisation (WHO). *Road Traffic Injuries*. WHO, Washington, D.C., 2016. www.who.int/mediacentre/factsheets/fs358/en/. Accessed July 31, 2017.

2. Bureau of Infrastructure, Transport and Regional Economics (BITRE). *Road Deaths Australia: 2013 Statistical Summary*. Department of Infrastructure and Regional Development, Canberra, 2016.
3. Department of Transport and Main Roads. *Queensland Road Toll Report No: 1021, Comparative Queensland Road Toll Year to Date to 23 July 2017*. www.tmr.qld.gov.au/Safety/Transport-and-road-statistics/Road-safety-statistics.aspx. Accessed July 23, 2017.
4. Department of Transport and Main Roads (DTMR). *Current Queensland Drivers' Licences*. DTMR, Brisbane, 2016.
5. Scott-Parker, B., and K. Rune. Review of the Graduated Driver Licensing Programs in Australasia. *Journal of the Australasian College of Road Safety*, Vol. 27, No. 4, 2016, pp. 15–22.
6. Scott-Parker, B., L. Bates, B. C. Watson, M. J. King, and M. K. Hyde. The Impact of Changes to the Graduated Driver Licensing Program in Queensland, Australia on the Experiences of Learner Drivers. *Accident Analysis & Prevention*, Vol. 43, No. 4, 2011, pp. 1301–1308.
7. Bates, L., B. Watson, and M. King. Required Hours of Practice for Learner Drivers: A Comparison Between Two Australian Jurisdictions. *Journal of Safety Research*, Vol. 41, No. 2, 2010, pp. 93–97.
8. Scott-Parker, B. Experiences of Young Drivers and Their advice for the Learner Licence Phase. *Traffic Injury Prevention*, Vol. 16, No. 2, 2015, pp. 109–115.
9. Naz, S., and B. Scott-Parker. Obstacles to Engaging in Young Driver Licensing: Perspectives of Parents. *Accident Analysis and Prevention*, Vol. 99, Pt. A, 2017, pp. 312–320.
10. Vlakveld, W. *Hazard Anticipation of Young Novice Drivers*. SWOV, The Netherlands, 2011.
11. Endsley, M. R. Toward a Theory of Situation Awareness in Dynamic Systems. *Human Factors: The Journal of the Human Factors and Ergonomics Society*. Vol. 37, No. 1, 1995, pp. 32–64.
12. Crundall, D. Hazard Prediction Discriminates Between Novice and Experienced Drivers. *Accident Analysis & Prevention*, Vol. 86, 2016, pp. 47–58.
13. Catchpole, J., W. Macdonald, and P. Cairney. Skill Deficits and Motivational Factors in Accidents Involving Young Drivers. *Young Drivers, Seminar, 1997*, Sydney, New South Wales, Australia, 1998.
14. Deery, H. A. Hazard and Risk Perception among Young Novice Drivers. *Journal of Safety Research*, Vol. 30, No. 4, 2000, pp. 225–236.
15. Scott-Parker, B., L. Wilks, and W. Griffin. An Innovative Approach to Improving Young Driver Situation Awareness Pre-Licensure: A Simulator Study. *Proc. Road Safety and Simulation Conference*, Den Haag, 17–19 October 2017.
16. Scott-Parker, B., J. Caldwell, C. Jones, and N. Hansen. Are Parents and Learner Drivers Seeing the Same Road? Presented at 95th Annual Meeting of the Transportation Research Board, Washington, D.C., 2016.
17. Scott-Parker, B., and J. Caldwell. Novel Investigation of Night-Time Driving Perceptual Similarities and Differences in Learner Drivers and Their Parent Supervisors. Presented at 96th Annual Meeting of the Transportation Research Board, Washington, D.C., 2017.
18. Walker, G., N. Stanton, and P. Salmon. Cognitive Compatibility of Motorcyclists and Car Drivers. *Accident Analysis and Prevention*, Vol. 43, No. 3, 2011, pp. 878–888.
19. Scott-Parker, B., B. Watson, M. J. King, and M. K. Hyde. "They're Lunatics on the Road": Exploring the Normative Influences of Parents, Friends, and Police on Young Novice's Risky Driving Decisions. *Safety Science*, Vol. 50, No. 9, 2012, pp. 1917–1928.
20. Scott-Parker, B., B. Watson, M. J. King, and M. K. Hyde. "I Would Have Lost the Respect of Friends and Family if I Bent the Road Rules": Parents, Peers, and the Perilous Behaviour of Young Drivers. *Transportation Research Part F: Traffic Psychology and Behaviour*, Vol. 28, 2015, pp. 1–13.
21. Scott-Parker, B. (in press). "You're so Used to Having Someone Tell You What to Do": Experiences of Young Drivers during the Provisional Licence Phase. Transportation Research Part F.
22. Scott-Parker, B., B. Watson, M. J. King, and M. K. Hyde. Young, Inexperienced and on the Road: Do Novice Drivers Comply with Road Rules? *Transportation Research Record: Journal of the Transportation Research Board*, 2012. 2318: 98–106.
23. Goodwin, A. H., R. D. Foss, L. H. Margolis, and S. Harrell. Parent Comments and Instruction during the First Four Months of Supervised Driving: An Opportunity Missed? *Accident Analysis & Prevention*, Vol. 69, 2014, pp. 15–22.
24. Zakrajsek, M. S., J. T. Shope, A. I. Greenspan, J. Wang, C. R. Bingham, and B. G. Simons-Morton. Effectiveness of a Brief Parent-Directed Teen Driver Safety Intervention (Checkpoints) Delivered by Driver Education Instructors. *Journal of Adolescent Health*, Vol. 53, No. 1, 2013, pp. 27–33.
25. Mirman, J. H., A. E. Curry, F. K. Winston, W. Wang, M. R. Elliott, M. T. Schultheis, M. C. Fisher Thiel, and D. R. Durbin. Effect of the Teen Driving Plan on the Driving Performance of Teenagers before Licensure: A Randomised Clinical Trial. *JAMA Pediatrics*, Vol. 168, No. 8, 2014, pp. 764–771.
26. Ramirez, M., J. Yang, T. Young, L. Roth, A. Garinger, and L. Snetselaar. Implementation Evaluation of Steering Teens Safe: Engaging Parents to Deliver a New Parent-Based Teen Driving Intervention to Their Teens. *Health Education & Behavior*, Vol. 40, No. 4, 2013, pp. 426–434.
27. McKenna, F. P., M. S. Horswill, and J. L. Alexander. Does Anticipation Training Affect Drivers' Risk Taking? *Journal of Experimental Psychology: Applied*, Vol. 12, No. 1, 2006, pp. 1–10.
28. McDonald, C. C., A. H. Goodwin, A. K. Pradhan, M. R. Romoser, and A. F. Williams. A Review of Hazard Anticipation Training Programs for Young Drivers. *Journal of Adolescent Health*, Vol. 57, No. 10, 2015, pp. S15–S23.
29. Pradhan, A. K., A. Pollatsek, M. Knodler, and D. L. Fisher. Can Younger Drivers Be Trained to Scan for Information That Will Reduce Their Risk in Roadway Traffic Scenarios That are Hard to Identify as Hazardous? *Ergonomics*, Vol. 52, No. 6, 2009, pp. 657–673.
30. Crundall, D., B. Andrews, E. van Loon, and P. Chapman. Commentary Training Improves Responsiveness to Hazards in a Driving Simulator. *Accident Analysis & Prevention*, Vol. 42, No. 6, 2010, pp. 2117–2124.
31. Ehsani, J. P., C. R. Bingham, and J. T. Shope. Graduated Driver Licensing for New Drivers: Effects of Three States' Policies on Crash Rates among Teenagers. *American Journal of Preventive Medicine*, Vol. 45, No. 1, 2013, pp. 9–18.
32. Scott-Parker, B. *SAFER: An Innovative Approach to Improving Young Driver Situation Awareness Pre-Licensure: Situation*

Awareness Fast-Tracking, Including Identifying Escape Routes. Intervention in Progress. MAIC, Brisbane Australia, 2017.

33. Scott-Parker, B., B. Wilks, and B. Huang (under review). *Teen Perceptions Regarding a Parent-Focused Road Safety Intervention: A Brief Report*.
34. Scott-Parker, B., J. Caldwell, C. Jones, and N. Hansen. Are Parents and Learner Drivers Seeing the Same Road? Presented at 95th Annual Meeting of the Transportation Research Board, Washington, D.C., 2016.
35. Scott-Parker, B., and J. Caldwell. Novel Investigation of Night-Time Driving Perceptual Similarities and Differences in Learner Drivers and Their Parent Supervisors. Presented at 96th Annual Meeting of the Transportation Research Board, Washington, D.C., 2017.
36. Rakotinirainy, A., S. Demmel, A. Watson, M. M. Haque, J. Fleiter, and S. Washington. *Prevalence and Perception of Following Too Closely in Queensland*. Queensland, Centre for Accident Research & Road Safety – Queensland (CARRS-Q), 2017.
37. Scott-Parker, B. J. (under review). *SAFER 2.0: Revolutionising Road Risks for Young Novice Drivers by Addressing Situation Awareness, Exposure, and Emotion-Related Driving Risks*.
38. Silvi, C., B. Scott-Parker, and C. Jones (under review). *A Vulnerable Young Driver Population: An Innovative Approach to Conceptualising the Driving and Licensure Experiences and Road Safety Risk of Young Drivers with an ASD*.
39. Scott-Parker, B. (under review). *A Study of Situation Awareness Fast-Tracking, Including Identifying Escape Routes (SAFER) with Senior Drivers: SAFER-Senior*.

The Standing Committee on Operator Education and Regulation (ANB30) peer-reviewed this paper (18-03502).

Article

TRR
JOURNAL OF THE TRANSPORTATION RESEARCH BOARD

Transportation Research Record
2018, Vol. 2672(33) 11–19

DOI: 10.1177/0361198118768530
journals.sagepub.com/home/trr
SAGE

Latent Hazard Anticipation in Young Drivers: Review and Meta-Analysis of Training Studies

James Unverricht[1], Siby Samuel[2], and Yusuke Yamani[1]

Abstract

Young drivers are overrepresented in motor vehicle crashes, and are shown to be poorer at anticipating potential threats on the roadway compared with their more experienced peers. Literature demonstrates the effectiveness of driver training programs at improving young drivers' latent hazard anticipation performance. Various hazard anticipation training studies have been undertaken on different population demographics using different training scenario presentation modes and multiple evaluation testbeds. These error-based feedback training programs (3M) allow trainees to make a mistake, show them how to mediate the mistake, and provide an opportunity to master the target skills. The current meta-analytical review focused on 19 peer-reviewed training studies that utilized eye movements to measure improvements in drivers' latent hazard anticipation performance following training. The role of four moderating factors (mode of delivery – PC-based or non PC-based; presentation of training – egocentric or exocentric; method of evaluation – on-road or driving simulator; and age of sample – teen novices aged 16–17 or young drivers aged 18–21) on the training effects were explored. Overall, the current meta-analysis suggest that: (a) superficial improvements in training programs does not necessarily further improve the drivers' latent hazard anticipation; (b) drivers who completed a training program with both egocentric and exocentric training views achieved greater levels of latent hazard anticipation performance than those who completed a training program that contained either view, but not both; and (c) the effect sizes of the 3M-based training programs on latent hazard anticipation were greater for drivers aged 18–21 years than drivers aged 16–17.

Young drivers are overrepresented in motor vehicle crashes, particularly crashes involving intersections, rear-end crashes, and run-off-the-road crashes (*1*). It has been widely demonstrated that a major cause of novice teen-driver crashes is their failure to scan effectively for potential risks (*1– 4*). Although several PC-based training programs have been developed to address this problem, only a handful have been evaluated (*5–8*), and even fewer have been extensively evaluated on-road (*9*). However, all evaluations of the PC-based training programs indicate that such training may help novice drivers perform better, both on a driving simulator and on the open road. In this paper, we are interested in one specific driving skill shown to be poorer among novice drivers compared with experienced drivers – latent hazard anticipation (*10*).

Latent Hazard Anticipation: Definition and Measurement

Pradhan et al. showed that the young, novice, and older experienced drivers differed in their ability to anticipate latent hazards (*11*). The driving simulator study included 72 participants (24 novice, 24 young and 24 older, experienced drivers). Overall, only 25.8% of the novice drivers engaged in behaviors indicating risk recognition and anticipation compared with 40.1% of the younger experienced drivers and 69.6% of the older experienced drivers across all the evaluation scenarios.

Latent hazards are impending threats that may not have materialized on the roadway yet. For example, consider a two-lane roadway with a marked midblock crosswalk with a bus parked on the right lane in advance of the crosswalk, thereby blocking the drivers' view of a potential pedestrian at the right extremity of the crosswalk (Figure 1). Here, the drivers' ability to anticipate the presence of a potential pedestrian at the right extremity of the crosswalk indicates his/her ability to successfully anticipate a latent threat.

To measure hazard anticipation, predetermined launch and target zones are used to quantify drivers' glance into a

[1]Department of Psychology, Old Dominion University, Norfolk, VA
[2]Department of Systems Design Engineering, University of Waterloo, Canada

Corresponding Author:
Address correspondence to Siby Samuel: siby.samuel@uwaterloo.ca

Figure 1. Perspective view of latent hazard anticipation scenario with "launch zone" (blue area) and "target zone" (red area).

binary score of either "0" for the absence of glance landing on the target zone while in the launch zone, or "1" for the presence of glance at the target zone while in the launch zone. In brief, target zone may be defined as an area or region on the roadway where a driver needs to glance in order to anticipate a latent threat, whereas a launch zone is the area on the roadway when the driver should begin scanning toward the target zone to anticipate a latent threat.

Hazard Anticipation Training: Theoretical Desiderata and Training Approach

Several driver training programs have been developed and evaluated over the past 15 years to improve young drivers' latent hazard anticipation ability. These training programs have all demonstrated varying levels of effectiveness at improving both young, and novice drivers' ability to anticipate a potential hazard. These studies have differed in the mode of training (PC-based, tablet-based, or Internet-based), method of evaluation (driving simulator or on-road), demographic evaluated (16–17-year-old novice teen drivers, or 18–21-year-old young drivers), and the presentation of training scenarios (egocentric or exocentric), among several other characteristics. Some evaluations of the training have also included examination of long-term retention effects up to a year after training.

Notwithstanding the specific scenario type, the Risk Assessment and Perception Test (RAPT) training program, and every other latent hazard anticipation training program from its generational tree (described in more detail in the next section), has been designed based on empirical evidence in the literature of memory and learning. Briefly, the depth of processing involved in an event can influence the likelihood of recalling the event from long-term memory by creating stronger, long-lasting memory traces (*12, 13*). RAPT is also consistent with literature that indicates that near transfer is maximized when the cues required to retrieve the knowledge for transfer is directly present within the training (*14*). RAPT uses targeted training, or "hugging" (*15*), to maximize near transfer learning by providing trainees with explicit, top-down schematic views. Further, RAPT follows previous research that learning transfers are most likely when the principles required to generalize are explicit for the learner (*16, 17*). Specifically, RAPT not only presents a risky scenario, but also explains to the trainee the risk associated with the situation and shows the driver when such a risk is hidden and what areas of the roadway need to be fixated in the scenario.

The family of training programs targeted at improving latent hazard anticipation as defined by Fisher et al. (*10*) have all utilized an error-based feedback mechanism of training. Error feedback mechanisms have been successfully used to train behavior (*18, 19*). In the context of driving, Fisher and colleagues (*10*) have adapted the training mechanisms to incorporate three important aspects of training: mistakes, mediation, and mastery – sometimes referred to as the 3M approach in literature (*10*). The training programs are developed with the purpose of placing the driver in a situation in which they are afforded a chance to make a mistake, followed by providing the trainees with the appropriate mediation strategy when they commit mistakes, and an opportunity to master the mediation strategy.

Brief History of Hazard Anticipation Training: Generations of RAPT Training

This section discusses in brief, the evolution of successfully evaluated and reported latent hazard anticipation training programs. The training improvements are discussed in chronological order. The very first training program aimed at improving young drivers' hazard anticipation ability was the RAPT program (*20*). The RAPT program (RAPT1) was developed on a PC using top-down plan (exocentric) views of several scenarios known to elicit poor latent hazard anticipation behavior from young drivers. The training was evaluated on a driving simulator and shown to successfully improve the proportion of latent hazards anticipated by trainees. The trained drivers fixated the target zones 57.7% of the time, whereas the untrained drivers made such fixations only about 35.4% of the time – a percent point difference of 22.3% (*10, 20, 21*).

One of the potential weaknesses of the RAPT1 evaluation was that the trained novice drivers were evaluated on the driving simulator immediately after training, and therefore the retention of the training effectiveness was unexplored. Researchers used a version of RAPT (RAPT2) that incorporated perspective (egocentric) views of the training scenarios on the PC, in addition to the plan views from RAPT1, to examine the retention of training effectiveness. Simulator evaluation of RAPT2 again revealed a clear effectiveness of the training on hazard anticipation. The RAPT2-trained novice drivers correctly fixated the target zones 52.1% of the time compared with only 28.1% for untrained drivers, showing retention of training effectiveness (*20, 22*).

A more recent version of RAPT (RAPT3) has incorporated only perspective views and was also shown to be effective in open road evaluations in the town of Amherst, MA (*20*). More specifically, the trainee is presented with a set of sequentially advancing perspective snapshots of a scenario. The authors found that more trained drivers (70%) correctly fixated the target zones than untrained drivers (33%) – a difference similar to that observed on the driving simulator (*23*). However, this RAPT3 field evaluation was conducted using 18–21-year-old young drivers, and not novice teen drivers – the targeted population of drivers. A second evaluation of RAPT3 was conducted in the town of Greenfield, MA, which demonstrated long-term retention effects of RAPT3 on novice teen drivers compared with a cohort of placebo-trained teen drivers (*24*). In the field immediately after training, the RAPT3 group anticipated hazards 65.8% of the time compared with 61.9% 6 to 8 months after training, whereas the control group anticipated the hazards only 47.3% of the time immediately after training and 37.7% of the time 6 to 8 months later (*24*).

RAPT3 underwent several other follow-up enhancements including a commercially improved and enhanced version of hazard anticipation training called Road Aware^{TM}, based on RAPT, which had a professionally designed, superior training interface (Flash-based PC training) unlike previous versions of RAPT, which were all developed within a laboratory environment using applications such as Microsoft PowerPoint or Adobe Captivate (*25, 26*). RoadAware^{TM} was evaluated using a college-aged (18–22) driver population on a driving simulator, and shown to be effective at improving young drivers' hazard anticipation ability. Another commercially enhanced version of RAPT (SuperRAPT) was first evaluated on the open road to measure the effectiveness of training on hazard anticipation in a study involving over 200 drivers (*27*), and subsequently in a naturalistic evaluation on crashes spanning over 5,000 drivers in the state of California (*9*). SimRAPT, employing plan views and a short video, was another version of the RAPT training that was developed and delivered on a driving simulator platform (*28*). An evaluation of SimRAPT on a driving simulator showed improvements for the trained driver group on their hazard anticipation behavior. Specifically, in near transfer scenarios, trained drivers were found to fixate at the hazardous region about 84% of the time compared with only 57% of untrained drivers (*28*).

A tablet version of the Road Aware^{TM} training program (Engaged Driver Training System, EDTS) was subsequently developed and customized to work on the iPad. The augmented training module –EDTS – was not only unique in its mode of delivery but was also additionally intended to train young drivers to reduce their engagement in secondary, distracting events in potentially hazardous situations [*29*]. The iPad provided participants with varying levels of control to navigate the virtual training scenarios. Participants are afforded control of the steering (by tilting the iPad from 0 to 90°), acceleration/braking (by toggling a button upwards/downwards), and scanning the road through 180° (by toggling a button toward the left/right). The EDTS training program was first evaluated on a driving simulator (*29*) and subsequently in the field (*30*) in two separate experiments, once immediately following training, and in an on-road evaluation, 7 to 8 months following training (*31*). All evaluations of the EDTS training reported significant training effects on drivers' hazard anticipation behavior compared with control conditions.

More recently, several other integrated training programs (e.g., SAFE-T, and ACCEL) have been developed that simultaneously train hazard mitigation (*32*) and attention maintenance skills (*33*) in addition to hazard anticipation behavior (*20*). Simulator evaluation of SAFE-T (a training program that integrates training on hazard anticipation, hazard mitigation, and attention maintenance skills) demonstrated effectiveness of training for trained young drivers compared with control (*34*), and RAPT training conditions (*35*), even though the RAPT portion of the program took only one-third of the duration of the original RAPT program. SAFE-T was also administered to improve the hazard anticipation behavior of individuals following either a short or a long total-awake-time, and was shown to be effective for long shift workers (*36*). ACCEL is an Internet-disseminable, omnibus training program that takes SAFE-T a step further and trains on both the tactical and strategic components of hazard anticipation, hazard mitigation, and attention maintenance behaviors. The ACCEL training administered multiple doses of the training to investigate the effect of training dosage on the size of the training effect. Analyses of eye movement data from a simulator evaluation revealed robust effects of hazard anticipation training for the trained teen-driver cohorts compared with the untrained or placebo-trained cohort (*37, 38*).

Outline of a Typical 3M Hazard Anticipation Training Scenario

For better illustration of how the training progresses, one training scenario developed in RAPT3 is discussed briefly (the complete training program can be accessed at www.ecs.umass/hpl). RAPT3 had a total of nine training scenarios in which there was a risk of collision with a potential pedestrian or vehicle. As the perspective views had to be photographed on-road (in real-time traffic conditions) for the training program, only scenarios that did not directly involve a moving vehicle were used to capture the snapshots for the training. Some staging (for example, parking a car to create an obstruction) was necessary to capture the various elements across the nine scenarios (*32*). For example, in the hidden sidewalk scenario, a sequence of still photographs shows the driver approaching a stop-controlled intersection with a crosswalk ahead of the stop line. A high hedge on the right beyond the stop sign obscures the visibility of a sidewalk from the right that merges onto the crosswalk. The risk is that a pedestrian or bicyclist may suddenly enter the crosswalk from the hidden sidewalk. While setting up the test course for evaluation, this test intersection was studied by the authors (*32*). Of the 20 drivers that were observed, all drivers failed to both stop at the stop line and scan toward the right periphery as they navigated past the bushes. Instead, they directly passed over the crosswalk and onto the cross street (*32*).

RAPT3 training program begins with instructions, an initial practice section to familiarize participants with the displays and tasks to be performed, followed by three main sections of the training – pre-test, training, and post-test. In the pre-test, each of the nine training scenarios were presented as a sequence of 5 to 12 progressive snapshots, each snapshot displayed for 3 s. The snapshots display the drivers' perspective view when navigating the situation (a hidden sidewalk in the above scenario example). The snapshots were generally of straight ahead views of the car. However, in situations that required the driver to look toward the left or right (for example, at an intersection), click buttons were provided on the left/right margins of the snapshot that allowed the user to see the corresponding left or right views. Trainees used a mouse to click over areas of the snapshot they thought presented a potential risk to them and that they would pay attention to if they were navigating the scenario on the road. The click coordinates and response times were recorded. However, no form of performance feedback was provided to trainees on the pre-test.

The real training occurred next; the trainees were shown a top-down schematic view of the scenario, with accompanying descriptions and explanations on the risky aspects associated with that scenario. Following the explanation, the user was again presented with a snapshot of the perspective view for that scenario. Relevant responses (mouse clicks) that were positioned within the critical area (target zone) were marked as correct. The shape of the target zone varied across the scenarios but were generally rectangular, including a slight tolerance for inaccuracy with mouse positioning. The program moved to the next scenario if the user correctly marked the target area. In the case of incorrect answers, users were taken back to the top-down schematic views and explanations for that scenario. Users were provided up to four opportunities to correctly identify the risky areas in the snapshot.

The post-test section presented the user with sequences of snapshots involving the nine scenarios, just as in the pre-test. And like the pre-test, click coordinates and response times were recorded and no feedback was provided. This way, the trainee in RAPT3 makes errors, mitigates with feedback, and masters the anticipation skills in pre-test, training, and post-test, respectively.

Current Study: Motivation and Approach

Despite the plethora of studies that focus on the improvements in latent hazard anticipation following training, there are still several questions raised when researchers embark upon the development and assessment of a targeted skill training intervention. Specifically, there are questions about the optimal platform for training, optimal presentation of the training (egocentric or exocentric or both), appropriate measurement testbed for behavior (simulator or on-road), and even the population demographic (novice teen drivers or younger drivers) that may benefit the most from such training.

The rest of the current paper reports on a preliminary meta-analysis of several young- and teen-driver training experiments specifically focused on improving young and novice teen drivers' latent hazard anticipation ability. The meta-analysis focuses on reported experiments spanning 14 years' duration (2004–present), in which a post-training evaluation component was also administered. Using the technique introduced by Rosenthal and Dimatteo (*39*) and Horrey and Wickens (*40*), the current meta-analysis quantitatively combines effect sizes reported in the experiments and explores potential moderators that influence the effectiveness of the training programs (mode of delivery, evaluation method, sample population, presentation of training). Across all the studies, the dependent measure was the proportion of latent hazards anticipated, strictly following the original measure introduced by Fisher et al. (*10*). All experiments considered in the analysis include either novice teen drivers (16–17 years) or young drivers (18–25 years) or both. All the experiments were either conducted on a driving simulator or the open road, and the eyes of the participants

Table 1. A List of Studies Included in the Meta-Analysis and Attributes of Potential Moderators

Study	Evaluation type	Presentation mode	Driver age
Fisher et al. (*10*)	Simulator	Exocentric	Teen (16–17)
Fisher et al. (*20*)	Simulator	Both	Teen (16–17)
Fisher et al. (*20*)	On-Road	Egocentric	Young (18–21)
Fisher et al. (*20*)	On-Road	Egocentric	Young (18–21)
Pradhan et al. (*32*)	On-Road	Egocentric	Young (18–21)
Vlakveld et al. (*28*)	Simulator	Exocentric	Young (18–25)
Taylor et al. (*24*)	On-Road	Egocentric	Teen (16–17)
Samuel et al., Zafian et al. (*25, 26*)	Simulator	Egocentric	Young (18–22)
Yamani et al. (*34*)	Simulator	Both	Teen (16–18)
Krishnan et al. (*29*)	Simulator	Egocentric	Young (17–19)
Hamid et al. (*36*)	Simulator	Both	Young (18–25)
Yamani et al. (*35*)	Simulator	Both	Teen (16–17)
Zafian et al. (*30*)	On-Road	Egocentric	Teen (16–17)
Zafian et al. (*31*)	On-Road	Egocentric	Teen (16–17)
Young et al. (*37*)	Simulator	Both	Teen (16–17)
Yamani et al. (*42*)	Simulator	Both	Young (18–21)
Agrawal et al. (*43*)	Simulator	Exocentric	Young (18–21)
Agrawal et al. (*43*)	Simulator	Egocentric	Young (18–21)
Fisher et al. (*38*)	Simulator	Both	Teen (16–17)

Note: Yamani et al. (*34*) evaluated an integrated training program (SAFE-T), whereas Yamani et al. (*35*) additionally evaluated RAPT3.

were tracked and eye movements recorded throughout to examine the key dependent measure of interest (latent hazard anticipation). Further, there was a high level of consistency in the equipment used, procedures followed, and the experimental design employed across the various studies included in this meta-analysis. Although different studies utilized variations of the parent hazard anticipation training program, the training approach and the core desiderata (e.g., 3M training mechanism) were identical across all included experiments. The selection criteria for study, data considered, and other caveats are reported in greater detail in the Method section.

Method

Studies

We identified 19 reported (peer-reviewed, full-length conference proceedings and journal articles; Table 1) experiments that examined the effectiveness of driver training programs that employed the 3M error feedback training mechanism on latent hazard anticipation for both, young (18–25) and novice teen (16–17) drivers. These drivers consisted of high school students recruited via a town driving school, college students from the five-college consortium including Amherst, and community college students from the Hampshire and Franklin counties. All participants were remunerated for their participation. We excluded studies that measured the effectiveness of relevant training programs using dependent measures other than drivers' glances made toward the target zones. Where available, we coded the reported test statistic, df (degrees of freedom), measures of evidence (e.g., *p*-value), and the three potential moderators: evaluation type (simulator vs. on-road evaluation), presentation mode (exocentric vs. egocentric), and driver age (teen vs. young). All the identified studies used an eye tracker to collect and record participant eye movements, and the hazard anticipation data was binary scored (0 – if the participant fails to glance at the target zone while in the launch zone; 1 – if the participant glances appropriately at the target zone while traversing the launch zone), and therefore, binomially distributed.

Meta-Analysis

The current meta-analysis required three main steps: 1) conversion of statistical results into appropriate effect sizes, 2) combining the effect sizes across the various experiments, and 3) constructing confidence intervals (CIs) around meta-analytic means of interest. First, following Rosenthal and Dimatteo (*39*) and Horrey and Wickens (*40*), we converted the test statistics with associated degrees of freedom reported in each study into r_{ES} as a measure of effect size. Specifically, the r_{ES} can be calculated from the t statistic, and F statistic with df of 1 in the numerator, as shown by

$$r_{ES} = \sqrt{\frac{t^2}{t^2 + df}} = \sqrt{\frac{F}{F + df_{error}}} \quad (1)$$

Where applicable, effect sizes were computed from chi square or the standard normal deviate Z using

$$r_{ES} = \sqrt{\frac{\chi^2(1)}{N}} \quad (2a)$$

$$r_{ES} = \frac{Z}{\sqrt{N}} \quad (2b)$$

where Z is the standard normal deviate and N is the sample size.

For articles reporting only p-values, following Rosenthal and Dimatteo (*39*), we converted them to Zs by using the following rule: for $p < .05$, $Z = 1.645$; for $p < .01$, $Z = 2.362$, and for $p < .001$, $Z = 3.090$. Note that some chi square values did not produce calculable r_{ES} values and the p-values were used to estimate the effect sizes. Such operations only produce more conservative effect sizes than the original chi square values.

Second, in order to aggregate the effect sizes, we applied Fisher's r-to-z transformation to normalize the effect sizes, and then calculated the unweighted mean of the transformed z-scores. Each transformed z-score is re-weighted according to each df of the study, and the weighted mean of the z-scores are calculated. The transformed z-scores are transformed back to r_{ES} for interpretation.

Finally, we constructed 95% CIs for the unweighted and weighted means using Equation 3. Note that CIs that exclude zero imply statistical significance.

$$CI = \overline{Z}_r \pm t_{(.05)} \frac{S}{\sqrt{k}} \quad (3)$$

where

$\overline{Z}_r$ is the mean of the transformed z-scores from rES,

t (.05) is the critical t-value for the alpha level of .05,

S is the standard deviation of the transformed z-scores, and

k is the number of the studies included in the analysis.

Results and Discussion

Table 2 presents the weighted and unweighted means with the CIs for each group. Additionally, we examined the effects of the four potential moderators of the training effectiveness on latent hazard anticipation performance. The results indicate five key findings. First, and most notably, the hazard anticipation training programs that utilize the 3M error-based training approach improve young and novice teen drivers' latent hazard anticipation, and its effect is determined to be large based on the combined measure of effect sizes (*41*). Second, the meta-analytic means and CIs indicate that, overall, the effectiveness of the programs remains large and that variability in delivery methods for the programs (PC-based vs. not PC-based programs) do not necessarily lead to measurable differences in the drivers' latent hazard anticipation performance.

Table 2. Summary for the Meta-Analysis and Examination of Potential Moderating Variables: Unweighted and Weighted Means of Effect Sizes (r) and Associated CIs (z = Transformed Effect Sizes)

	Unweighted (95% CI)	Weighted (95% CI)
Overall (N = 19)	0.536 (.453, .745)	0.495 (.349, .641)
Mode of Delivery		
PC-Based (N = 12)	0.475 (.432, .603)	0.466 (.383, .628)
Not PC-Based (N = 7)	0.496 (.329, .759)	0.448 (.170, .795)
Evaluation Type		
Simulator (N = 13)	0.564 (.439, .839)	0.518 (.368, .780)
On-road (N = 6)	0.539 (.384, .821)	0.507 (.394, .725)
Presentation Mode		
Exocentric (N = 3)	0.489 (.020, 1.05)	0.441 (.072, .874)
Egocentric (N = 9)	0.536 (.459, .741)	0.421 (.32, .581)
Both (N = 7)	0.405 (.339, .522)	0.526 (.193, .977)
Driver Age		
Teen (N = 9)	0.408 (.344, .525)	0.412 (.207, .670)
Young (N = 10)	0.544 (.488, .734)	0.520 (.428, .727)

Researchers may strive to enhance quality of the interfaces of such programs, but the current results imply that the effects of usability of the programs on the participants' actual anticipation performance can be trivial. Another noteworthy point is that the mean latent hazard anticipation rate for the trained group is not close to the perfect anticipation performance (100%), even though many of the drivers [except for those tested for far transfer (*43*)] were primarily evaluated on scenarios that were identical to the scenarios they were trained on in terms of knowledge transfer. Superficial improvement on the training programs does not appear to necessarily further improve the drivers' latent hazard anticipation, and other auxiliary techniques should be explored such as using expert drivers' eye movements (*42*) or the use of virtual reality, (*43*) or both, in the training programs.

Third, for evaluation type, the meta-analytic means of the effect sizes did not appear to differ between the simulator and on-road evaluations. Note, though, that there were only six qualified studies included in the on-road evaluation group, making the estimates of the means and variability less reliable. Fourth, for presentation mode, though statistically unreliable, the weighted means show the training programs utilizing both egocentric and exocentric presentations of the training scenarios produced the largest effect size when compared with the egocentric-only and the exocentric-only presentations. This implies that the drivers who completed a training program with both top-down and driver's perspective views achieved greater levels of latent hazard anticipation performance than those who completed a training program that contained either view but not both. Related with the above, teen drivers may need additional opportunities to learn critical scenarios in both egocentric as well as exocentric perspectives to accelerate and maximize their learning. Furthermore, it remains unclear how trainees integrate different modes of spatial and dynamic information about driving scenarios. More research is clearly needed to investigate this issue.

Lastly, the current meta-analysis shows that the effect sizes of the 3M-based training programs on latent hazard anticipation were greater for young drivers aged 18–21 years than teen drivers aged 16–18. Although further accumulation of evidence is necessary for drawing a more definitive conclusion, this result is counterintuitive because commonly cited statistics (*44*) show that teen drivers are at an increased risk of fatal crashes per 100 million mi traveled than their young counterparts. That is, teen drivers are more likely to be poorer at anticipating latent hazards than young drivers mainly because of their lack of experience with critical driving scenarios (*1*). The current results imply that the young drivers benefit from the training programs more than teen drivers.

Speculatively, there are at least two distinct possibilities. First, with more driving experience since the licensure, young drivers might be able to learn efficiently from the training programs by relating the training materials to their direct experiences. It has been hypothesized that cognitive development in adolescence involves regulatory competence in emotion-related behavior such as risk-taking behaviors, and individuals in late adolescence become more able to control their affective responses to emotion-triggering events, signified by changes in their prefrontal cortex, controlling central executive functions (*45*). Second, it is possible that a larger effect size was obtained for young drivers because their pre-training baseline performance was poorer than teen drivers due to the cohort effect. However, we believe this account is unlikely because the studies were conducted across different years with different sets of driver populations. More studies should replicate the findings and examine the underlying mechanisms that produce a larger effect of training programs on latent hazard anticipation for young than teen drivers. The results confirm that the training programs have proven effective in improving latent hazard anticipation performance for not only teen drivers but also young drivers who do not have experience with the critical scenarios covered in the training programs, beyond the level of benefit as in, young drivers have been shown to gain more from training than teen drivers. Of note is the recent study conducted in California by Thomas et al., based on over 5,000 drivers randomly assigned to either the RAPT program or a placebo condition, that showed a treatment by gender interaction in which RAPT reduced crash frequency only for males (*9*). The study also showed significant improvements in scores on a hazard perception test, but no evidence that the improved scores were necessarily associated with crashes. This study was not included in the current meta-analysis. Other similar randomized clinical trials are currently underway to study the overall impact of hazard anticipation training on reducing crashes. Based on the trends observed in the meta-analysis, these large-scale studies are good investments for understanding the etiology of crashes.

Limitations and Conclusions

There are a few limitations with our approach. First, the number of studies used in the current meta-analysis is smaller than typically observed in other meta-analytical studies. However, the experiments used in the current analysis have a great level of consistency with the dependent measure, procedures, and design, and therefore offer more robustness to the sample set. Second, we examined four specific moderators of training that were easily identifiable from the reported literature; there may potentially be other moderators that could additionally explain the impact of training on improving hazard anticipation behavior. Third, other regression-modeling-based meta-analytical techniques may be applied to further glean information. However, given the statistical data reported, we used a relevant approach that had been previously utilized and validated in studies (*39, 40*).

In conclusion, the current meta-analysis confirmed the large effect size of 3M-based training programs on improving drivers' latent hazard anticipation performance. We further examined the four potential moderating variables: mode of delivery, evaluation type, presentation mode, and driver age. Though the number of studies included in the analysis was limited due to our strict selection criteria (e.g., employing the measure of hazard anticipation originally established by Fisher et al. (*10*), the data patterns suggest that, 1) drivers obtain a similar effect from a training program when evaluated in simulator or on-road, 2) training programs that present scenarios with both egocentric and exocentric spatial information tend to produce a larger training effect than those with either information type, and 3) latent hazard anticipation training may be of greater benefit for young- than teen-driver populations. The findings of this study offer valuable insight in terms of training study design considerations for other researchers interested in the development and evaluation of similar young-driver-targeted training programs. Further, the study has direct utility to government agencies for the assessment of the collective impact of training interventions on long-term improvements to driver behavior.

Acknowledgments

Portions of this research were supported by the Summer Faculty Research Program, Office of Research, Old Dominion University awarded to Y.Y., and a Graduate STEM Faculty Student Research grant from University of Massachusetts Amherst to S.S.

References

1. McKnight, A. J., and A. S. McKnight. Young Novice Drivers: Careless or Clueless? *Accident Analysis & Prevention*, Vol. 35, No. 6, 2003, pp. 921–925.
2. Fisher, D., and A. Pollatsek. Novice Driver Crashes: Failure to Divide Attention or Failure to Recognize Risks. *Attention: From Theory to Practice*, Vol. 134, 2007, p. 156.
3. Pradhan, A. K., K. R. Hammel, R. DeRamus, A. Pollatsek, D. A. Noyce, and D. L. Fisher. Using Eye Movements to Evaluate Effects of Driver Age on Risk Perception in a Driving Simulator. *Human Factors*, Vol. 47, No. 4, 2005, pp. 840–852.
4. Garay-Vega, L., and D. L. Fisher. Can Novice Drivers Recognize Foreshadowing Risks as Easily as Experienced Drivers. *Proc., 3rd International Driving Symposium on Human Factors in Driver Assessment, Training and Vehicle Design*, Rockport, Maine, 2005, pp. 471–477.
5. Allen, R. W., T. J. Rosenthal, G. Park, M. Cook, D. Fiorentino, and E. Viirre. Experience with a Low-Cost PC-Based System for Young Driver Training. *Proc., International Conference on Driver Behaviour and Training*, Stratford-Upon-Avon, 2003, pp. 359–370.
6. Willis, D. K. The Impetus for the Development of a New Risk Management Training Program for Young Teen Drivers. *Proceedings of the Human Factors and Ergonomics Society Annual Meeting*, Vol. 42, No. 20, 1998, pp. 1394–1395.
7. Fisher, D. L., N. E. Laurie, R. Glaser, K. Connerney, A. Pollatsek, S. A. Duffy, and J. Brock. The Use of an Advanced Driving Simulator to Evaluate the Effects of Training and Experience on Drivers' Behavior in Risky Traffic Scenarios. *Human Factors*, Vol. 44, No. 2, 2002, pp. 287–302.
8. Regan, M. A., H. A. Deery, and T. J. Triggs. Training for Attentional Control in Novice Car Drivers: A Simulator Study. *Proceedings of the Human Factors and Ergonomics Society Annual Meeting*, Vol. 42, No. 20, 1998, pp. 1452–1456.
9. Thomas, F. D., S. Rilea, R. D. Blomberg, R. C. Peck, and K. T. Korbelak. *Evaluation of the Safety Benefits of the Risk Awareness and Perception Training Program for Novice Teen Drivers*. Report No. DOT HS 812 235. National Highway Traffic Safety Administration, Washington, D.C., 2016.
10. Fisher, D. L., V. Narayanaan, A. Pradhan, and A. Pollatsek. Using Eye Movements in Driving Simulators to Evaluate Effects of PC-Based Risk Awareness Training. *Proceedings of the Human Factors and Ergonomics Society Annual Meeting*, Vol. 48, No. 19, 2004, pp. 2266–2270.
11. Pradhan, A. K., K. R. Hammel, R. DeRamus, A. Pollatsek, D. A. Noyce, and D. L. Fisher. Using Eye Movements to Evaluate Effects of Driver Age on Risk Perception in a Driving Simulator. *Human Factors*, Vol. 47, No. 4, 2005, pp. 840–852.
12. Craik, F. I., and R. S. Lockhart. Levels of Processing: A Framework for Memory Research. *Journal of Verbal Learning and Verbal Behavior*, Vol. 11, No. 6, 1972, pp. 671–684.
13. Craik, F. I. Levels of Processing: Past, Present … and Future? *Memory*, Vol. 10, No. 5–6, 2002, pp. 305–318.
14. Salomon, G., and D. N. Perkins. Rocky Roads to Transfer: Rethinking Mechanism of a Neglected Phenomenon. *Educational Psychologist*, Vol. 24, No. 2, 1989, pp. 113–142.
15. Perkins, D. N., and G. Salomon. Teaching for Transfer. *Educational Leadership*, Vol. 46, No. 1, 1988, pp. 22–32.
16. Gick, M. L., and K. J. Holyoak. Analogical Problem Solving. *Cognitive Psychology*, Vol. 12, No. 3, 1980, pp. 306–355.
17. Gick, M. L., and K. J. Holyoak. Schema Induction and Analogical Transfer. *Cognitive Psychology*, Vol. 15, No. 1, 1983, pp. 1–38.
18. Gist, M. E., C. Schwoerer, and B. Rosen. Effects of Alternative Training Methods on Self-Efficacy and Performance in Computer Software Training. *Journal of Applied Psychology*, Vol. 74, No. 6, 1989, p. 884.
19. Ivancic, K., and B. Hesketh. Learning from Errors in a Driving Simulation: Effects on Driving Skill and Self-Confidence. *Ergonomics*, Vol. 43, No. 12, 2000, pp. 1966–1984.
20. Fisher, D. L., A. P. Pollatsek, and A. Pradhan. Can Novice Drivers be Trained to Scan for Information that will Reduce Their Likelihood of a Crash? *Injury Prevention*, Vol. 12, 2006, pp. i25–i29.
21. Pollatsek, A., V. Narayanaan, A. Pradhan, and D. L. Fisher. The use of Eye Movements to Evaluate the Effect of PC-Based Risk Awareness Training on an Advanced Driving Simulator. *Human Factors*, Vol. 48, No. 3, 2006, pp. 447–464.
22. Pradhan, A., D. Fisher, and A. Pollatsek. Risk Perception Training for Novice Drivers: Evaluating Duration of Effects of Training on a Driving Simulator. *Transportation Research Record: Journal of the Transportation Research Board*, 2006. 1969: 58–64.
23. Pradhan, A. K., D. L. Fisher, A. Pollatsek, M. Knodler, and M. Langone. Field Evaluation of a Risk Awareness and Perception Training Program for Younger Drivers. *Proceedings of the Human Factors and Ergonomics Society Annual Meeting*, Vol. 50, No. 22, 2006, pp. 2388–2391.
24. Taylor, T. G., K. M. Masserang, A. K. Pradhan, G. Divekar, S. Samuel, J. W. Muttart, A. Pollatsek, and D. L. Fisher. Long Term Effects of Hazard Anticipation Training on Novice Drivers Measured on the Open Road. *Proc., 6th International Driving Symposium on Human Factors in Driver Assessment, Training, and Vehicle Design*, Olympic Valley, Lake Tahoe, Calif., 2011, pp. 187–194.
25. Samuel, S., T. Zafian, A. Borowsky, M. R. Romoser, and D. L. Fisher. Can Young Drivers Learn to Anticipate Hidden Hazards: A Driving Simulator Study. *Proc., 7th International Driving Symposium on Human Factors in Driver Assessment, Training, and Vehicle Design*, Bolton Landing, New York, Public Policy Center, University of Iowa, 2013, pp. 355–361.
26. Zafian, T., S. Samuel, A. Borowsky, and D. L. Fisher. *Can Young Drivers be Trained to Better Anticipate Hazards in Complex Driving Scenarios? A Driving Simulator Study.* Presented at 93rd Annual Meeting of the Transportation Research Board, Washington, D.C., 2014.
27. Divekar, G., S. Samuel, F. D. Thomas, K. Korbelak, R. D. Blomberg, and D. L. Fisher. On-Road Evaluation of Hazard Perception Training Program Comparing Eye Tracking and Commentary Driving Techniques. Presented at 95th Annual Meeting of the Transportation Research Board, Washington, D.C., 2016.

28. Vlakveld, W., M. Romoser, H. Mehranian, F. Diete, A. Pollatsek, and D. Fisher. Do Crashes and Near Crashes in Simulator-Based Training Enhance Novice Drivers' Visual Search for Latent Hazards? *Transportation Research Record: Journal of the Transportation Research Board*, 2011. 2265: 153–160.
29. Krishnan, A., S. Samuel, M. Romoser, and D. Fisher. A Study to Evaluate the Effectiveness of a Tablet-Based Training Program for Younger Drivers: Addressing Distraction Head On. *Proceedings of the Human Factors and Ergonomics Society Annual Meeting*, Vol. 59, No. 1, 2015, pp. 1651–1655.
30. Zafian, T. M., S. Samuel, J. Coppola, E. G. O'Neill, M. R. Romoser, and D. L. Fisher. On-Road Effectiveness of a Tablet-Based Teen Driver Training Intervention. *Proceedings of the Human Factors and Ergonomics Society Annual Meeting*, Vol. 60, No. 1, 2016, pp. 1926–1930.
31. Zafian, T. M., R. Agrawal, and S. Samuel. *On-Road Assessment of Retention Effects on Hazard Anticipation Training for Novice Drivers*. Presented at 96th Annual Meeting of the Transportation Research Board, Washington, D.C., 2017.
32. Pradhan, A. K., A. Pollatsek, M. Knodler, and D. L. Fisher. Can Younger Drivers be Trained to Scan for Information that will Reduce Their Risk in Roadway Traffic Scenarios that Are Hard to Identify as Hazardous? *Ergonomics*, Vol. 52, No. 6, 2009, pp. 657–673.
33. Pradhan, A. K., G. Divekar, K. Masserang, M. Romoser, T. Zafian, R. D. Blomberg, F. D. Thomas, I. Reagan, M. Knodler, A. Pollatsek, and D. L. Fisher. The Effects of Focused Attention Training on the Duration of Novice Drivers' Glances Inside the Vehicle. *Ergonomics*, Vol. 54, No. 10, 2011, pp. 917–931.
34. Yamani, Y., S. Samuel, and D. Fisher. Simulator Evaluation of an Integrated Road Safety Training Program. *Proceedings of the Human Factors and Ergonomics Society Annual Meeting*, Vol. 58, No. 1, 2014, pp. 1904–1908.
35. Yamani, Y., S. Samuel, M. A. Knodler, and D. L. Fisher. Evaluation of the Effectiveness of a Multi-Skill Program for Training Younger Drivers on Higher Cognitive Skills. *Applied Ergonomics*, Vol. 52, 2016, pp. 135–141.
36. Hamid, M., S. Samuel, A. Borowsky, W. J. Horrey, and D. L. Fisher. Evaluation of Training Interventions to Mitigate Effects of Fatigue and Sleepiness on Driving Performance. *Transportation Research Record: Journal of the Transportation Research Board*, 2016. 2584: 30–38.
37. Young, J., T. Zhang, S. Samuel, M. Knodler, and D. L. Fisher. Simulator Evaluation of the Effectiveness of an Comprehensive Teen Driver Training Program. Presented at 96th Annual Meeting of the Transportation Research Board, Washington, D.C., 2017.
38. Fisher, D. L., J. Young, T. Zhang, M. Knodler, and S. Samuel. *Accelerating Teen Driver Learning: Anywhere, Anytime Training*. Final Report. AAA Foundation for Traffic Safety, Washington, D.C., 2017.
39. Rosenthal, R., and M. R. DiMatteo. Meta-Analysis: Recent Developments in Quantitative Methods for Literature Reviews. *Annual Review of Psychology*, Vol. 52, No. 1, 2001, pp. 59–82.
40. Horrey, W. J., and C. D. Wickens. Examining the Impact of Cell Phone Conversations on Driving Using Meta-Analytic Techniques. *Human Factors*, Vol. 48, No. 1, 2006, pp. 196–205.
41. Cohen, J. *Statistical Power Analysis for the Behavioral Sciences*. Lawrence Earlbaum Associates, Hilsdale, N.J., 1988.
42. Yamani, Y., P. Bıçaksız, D. B. Palmer, J. M. Cronauer, and S. Samuel. Following Expert's Eyes: Evaluation of the Effectiveness of a Gaze-Based Training Intervention on Young Drivers' Latent Hazard Anticipation Skills. *Proc., 9th International Driving Symposium on Human Factors in Driver Assessment, Training, and Vehicle Design*, Public Policy Center, University of Iowa, Manchester Village, Vermont, 2017, pp. 129–135.
43. Agrawal, R., M. Knodler, D. Fisher, and S. Samuel. Advanced Virtual Reality Based Training to Improve Young Drivers' Latent Hazard Anticipation Ability. *Proceedings of the Human Factors and Ergonomics Society Annual Meeting*, Vol. 61, No. 1, 2017, pp. 1995–1999.
44. Insurance Institute for Highway Safety. *Analysis of Data from the Fatality Analysis Reporting System*. Arlington, Va., 2008.
45. Steinberg, L. Cognitive and Affective Development in Adolescence. *Trends in Cognitive Sciences*, Vol. 9, 2005, pp. 69–74.

The Standing Committee on Operator Education and Regulation (ANB30) peer-reviewed this paper (18-02056).

Article

TRR
JOURNAL OF THE TRANSPORTATION RESEARCH BOARD

Transportation Research Record
2018, Vol. 2672(33) 20–30

DOI: 10.1177/0361198118758311
journals.sagepub.com/home/trr
SAGE

Virtual Reality Headset Training: Can It Be Used to Improve Young Drivers' Latent Hazard Anticipation and Mitigation Skills

Ravi Agrawal[1], Michael Knodler[1], Donald L. Fisher[2], and Siby Samuel[3]

Abstract

Young drivers are overrepresented in motor-vehicle crashes compared to experienced drivers. Research shows that young drivers are generally clueless, not careless, failing to anticipate and mitigate latent hazards. There are several error-feedback training interventions that emphasize the teaching of latent hazard anticipation skills (e.g., risk awareness and perception training, RAPT) and a few that emphasize both the teaching of hazard anticipation and hazard mitigation skills (e.g., the driver simulation ACCEL). In the current study, a virtual reality, headset-based latent hazard anticipation and mitigation training program (V-RAPT) was developed on a head-mounted display (Oculus Rift). The headset provides the participant with a 100-degree wide field of view of six high-risk driving scenarios, the view changing appropriately as the participant rotates his or her head. Thirty-six young drivers were exposed to one of three training programs—V-RAPT, RAPT, and a placebo—and then evaluated on a driving simulator. Eye movement and vehicle data were collected throughout the simulator evaluation. The drives included the six scenarios used in training and four other scenarios dissimilar to the ones used in training, but previously validated as measures of hazard anticipation. The drivers trained on V-RAPT were found to anticipate a significantly greater proportion (86.25%) of latent hazards than the RAPT (62.36%) and placebo (30.97%) trained drivers. The V-RAPT trained drivers were also found to be better at mitigating potential threats. The virtual reality, headset-based training program holds out the promise of improving drivers' ability to anticipate and mitigate latent threats and thereby reduce crashes.

Young drivers are at a greater risk of being involved in crashes than any other age group of drivers (*1*). The National Highway Traffic Safety Administration states that younger drivers' lack of experience, increased willingness to take risks, and greater immaturity are the primary reasons for this over-involvement in crashes (*2*). McKnight and McKnight reviewed over 1,000 crashes and identified inexperience as a major contributor toward crashes among young drivers (*3*). Specifically, failures to scan for the hazard ahead, to the front, the side, and the rear contributed to nearly 42.7% of the crashes in which young drivers are involved.

Previous driving simulator-based studies examining young drivers' hazard anticipation skills also suggest inexperience as a major factor, in which younger drivers fail to recognize both hazards that are visible and materialize as a threat at the very last second and hazards that are not visible, but are likely to materialize (*4*). The latter are typically referred to as latent threats. Latent threat hazard anticipation may be best explained by an example scenario (*5, 6*) Imagine a two-lane roadway (one travel lane in either direction) with a truck that is stopped on the right-hand side in a parking lane in front of a marked mid-block crosswalk. A driver is approaching the crosswalk. The truck is blocking the driver's view of a potential pedestrian on the crosswalk. Therefore, to safely navigate, the driver should scan toward the right front edge of the truck to identify the presence of any potential pedestrians. The results from a simulator study by Pradhan et al. demonstrate that experienced drivers (57.6%) were more likely to scan toward the right front edge than were younger drivers (28.6%) (*5*). The experienced

[1]Department of Mechanical and Industrial Engineering, University of Massachusetts Amherst, Amherst, MA
[2]Volpe National Transportation Systems Center, Cambridge, MA
[3]Department of Systems Design Engineering, University of Waterloo, Waterloo, ON, Canada

Corresponding Author:
Address correspondence to Ravi Agrawal: raviagrawal@umass.edu

drivers were also better at mitigating hazards as they were more likely to steer further toward the left as they passed in front of the truck (*6*).

Various training programs have been developed in recent years to improve young drivers' latent hazard anticipation ability and many of these programs have been shown to be effective through driving simulator studies and on-road evaluations (*7–14*). Risk awareness and perception training (RAPT) was one of the first driver training programs developed to address novice drivers' failure to appropriately scan and detect latent threats present/emerging on the forward roadway. The initial version of the training program was developed on a PowerPoint platform, delivered on a PC, and evaluated on a driving simulator. Subsequent versions of RAPT have been developed on advanced platforms such as smartphones, tablets, and driving simulators. Driving simulator and on-road assessments of the RAPT training program [and all its subsequent versions: RAPT3 (*15*), Distractology 101 (*16*), SuperRAPT (*17*), CalRAPT (*18*), and SimRAPT (*19*)] have shown that not only can RAPT increase hazard anticipation skills as measured on a driving simulator and in the field (on-road) (*5, 6, 9, 10, 20*), but it is the first program in some 50 years to show an actual reduction in crashes. The reduction in crashes in both studies (*17, 19*) was about 20%.

Training programs have also been designed to address both latent hazard anticipation and its subsequent mitigation (*21, 22*). Hazard anticipation is a necessary precursor of hazard mitigation and so it is always taught along with hazard mitigation (although the converse is not necessarily true). Referring to the scenario described above when the driver's view of the pedestrian entering the crosswalk is obstructed by a parked truck, to successfully mitigate the latent hazard a driver should slow down when passing the parked vehicle and move slightly to the right. Evaluations of training programs targeting hazard anticipation and hazard mitigation have shown clear benefits, improving both skills when the effects are evaluated on a driving simulator [e.g., Driver ZED, (5); ACCEL, (*23*)]. No training programs targeting both hazard anticipation and hazard mitigation have been evaluated in the field yet, using either vehicle behaviors or actual crashes as the dependent measures.

In summary, training programs targeting hazard anticipation have shown improvements in safe behaviors (glances toward the latent hazard) and decreases in crashes. Training programs targeting both hazard anticipation and hazard mitigation have shown, along with improvements in hazard anticipation, improvements in hazard mitigation. Curiously, most of the training programs targeting hazard anticipation have fallen short of getting novice drivers to the point when they anticipate the great majority of the hazards. For example, RAPT trained drivers correctly anticipate threats about 60% of the time compared to their placebo trained peers (a 30 percentage point gain). Although this is a doubling in performance among the trained drivers, the performance of the trained drivers is still nowhere near ceiling. There is a need to identify strategies of training that can further improve hazard anticipation performance while not detracting from hazard mitigation performance.

To understand what might be done additionally with hazard anticipation training, one needs to understand the current training methods. This paper focuses on training programs that utilize what is called error training in some contexts (*24, 25*) or 3M training in other contexts (*5, 6*). Specifically, 3M training includes three critical modules: a mistake, a mediation and a mastery module. In the mistake module, participants are presented with the opportunity to make a mistake (in which a mistake is a failure to anticipate a latent threat). In the mediation module, an explanation is provided of why a potential threat is not visible and what the participant should do about their behaviors. In the mastery module, participants are provided with the opportunity to master the correct behaviors which were explained in the mediation module. This training approach (error training or 3M training) has been successfully validated in several previous training studies (*9–12, 20, 25*).

The question is how could one enhance the hazard anticipation training and still maintain the same level of hazard mitigation training that other programs have achieved. Assuming that one could record an individual participant's trip through a virtual world from the participant's perspective, including the participant's eye glances (or head movements), and speed, one has all the data one needs to identify the effects of hazard anticipation and hazard mitigation training. Additionally, assuming one can play back an individual's trip through the virtual world in real time, one can then enhance the mediation module described above. In particular, the participant could see where he or she did not look and, as the latent threat was passed, reveal to the participant the threat which was hidden and did not materialize. This means that the errors become truly salient events and can motivate further learning (*26*). None of the current training programs provide this level of salience. Usually, the reason for the error is explained during the mediation module. But, the participant is not taken back through a drive, shown where he or she made an error, and shown the latent threat which could possibly have materialized.

The above enhancement to 3M training could be made in any virtual reality environment. However, the aim was to do it in an environment that was immersive, could be potentially broadly disseminated, and had a proven track record. Virtual reality headsets (VRHs) represent a new technology with just this possibility. They have been successfully utilized as a training platform in other domains (*27–31*). They are relatively inexpensive compared to even the least expensive of driving simulators by almost at least two orders of magnitude. They can be widely disseminated. They can be used to create immersive worlds through which a participant can travel and in which the participant's head movements (hazard anticipation) and speed (hazard mitigation) can be recorded. Moreover, a participant's actual trip through the virtual world can easily be recorded and replayed. Therefore, it was decided to explore the use of VRHs as a training tool for hazard anticipation and hazard mitigation.

It was investigated whether the drivers trained using the VRH-based risk awareness and perception training (V-RAPT) program would be able to anticipate a greater proportion of latent hazards compared to the drivers trained using RAPT (or the placebo). Further, because V-RAPT also trains the driver to mitigate the latent hazard following successful anticipation of the hazard, one seeks to compare whether the V-RAPT trained drivers are significantly better in their ability to mitigate hazards compared to their RAPT or placebo trained peers.

Method

Thirty-six young drivers were randomly and equivalently assigned to one of three training conditions (12 in each group). One group of young drivers was trained on V-RAPT, a second group of young drivers was trained using the PC-based RAPT program, and the last group received a PC-based placebo training program. The scenarios used to train hazard anticipation in V-RAPT and RAPT were conceptually identical to one another (i.e., the latent hazards were the same). The critical difference was the medium used to deliver the training (one a VRH through which the participant could navigate in real time, one a single screen of a PC with a static display). Following the training, all the participants were immediately evaluated on a driving simulator to identify the effectiveness of the training program using eye movements and speed as the dependent variables. Some of the hazard anticipation scenarios evaluated on the driving simulator were similar to those trained in V-RAPT and RAPT (referred to as *near transfer* scenarios); other hazard anticipation scenarios were quite dissimilar (referred to as *far transfer* scenarios).

Participants

Thirty-six subjects aged 18–25 were recruited for this study which had full approval from the University of Massachusetts Amherst Institutional Review Board. Data from one subject were excluded because of technical failures whereas two other participants dropped out from the study because of simulator sickness during the evaluation portion of the study (one V-RAPT and one RAPT participant). The 12 participants (seven male, five female) in the V-RAPT group had a mean age of 20.50 years (SD = 1.24) and a mean driving experience of 3.79 years (SD = 1.09). The 12 drivers (eight male, four female) in the RAPT training group had a mean age of 21.333 years (SD = 1.87) and mean driving experience of 3.63 years (SD = 1.99). The 12 drivers (eight male, four female) in the placebo training group had a mean age of 20.25 years (SD = 1.13) and mean driving experience of 3.43 years (SD = 1.81). There was no statistical difference among the ages of the drivers by training group or their months of licensure. All participants were recruited from the town of Amherst and surrounding areas and were remunerated for their participation in the study.

Apparatus

The current experiment utilized a driving simulator, an eye tracker, and a VRH (an Oculus Rift).

Driving Simulator. A Realtime Technologies Inc. full cab, fixed-base driving simulator was used to conduct this experiment. The simulator cab is a fully equipped 1995 Saturn sedan placed in front of three screens subtending 135 degrees horizontally and 180 degrees visually at a display resolution of 1240 × 768 pixels and at a frequency of 60 Hz. The participants control the simulator cab, just like a normal car. The vehicle dynamics, physics, and graphics provide a realistic experience to the participants. Realistic wind, Doppler, and environment sounds are generated through a surround system consisting of side speakers and two subwoofers located under the hood of the car.

Eye Tracker. An Applied Science Laboratories (ASL) Mobile Eye head mounted, portable, eye tracking unit was used to monitor and record fixations of the driver. The eye tracking unit is equipped with multiple cameras that produce scene and eye images which when superimposed, generates crosshairs of where the participants fixate on the roadway. The ASL eye tracker provides a visual angle range of 50 degrees in the horizontal direction and 40 degrees in the vertical direction with a system accuracy of approximately 0.5 degrees of visual angle.

Oculus Rift. An Oculus Rift is used in this study to deliver the V-RAPT training. The Oculus Rift comes with a VRH, motion sensor, remote, and an Xbox One wireless controller. The Rift has a rich organic light-emitting diode (OLED) screen with 1080 × 1200 resolution per eye and a refresh rate of 90 Hz. The Rift screen provides 100 degrees' field of view and the integrated headphones in the rift provide a three-dimensional audio effect. The motion sensing platform performs rotational and positional tracking using a USB stationary infrared sensor. With the use of the sensor, the Oculus Rift creates a virtual three-dimensional space where the user can sit, move, walk, or drive around.

Training

V-RAPT. The V-RAPT training program contains computer-generated videos displayed from the driver's eyepoint of six hazard anticipation scenarios that were used in RAPT (which had static, top-down views). The training program, described above, utilizes the enhanced 3M approach that is error and

Table 1. List of Virtual Reality Headset Scenarios with Perspective Views

Scenario	Scenario description	Perspective scenario views
Scenario 1: Curve + T Intersection	The driver is approaching a T-intersection where there is a vehicle waiting in the forward lane in the opposite direction and another vehicle on the connecting road (cross street). The vehicle on the cross street is blocking the view of a potential pedestrian. The driver needs to scan for the pedestrian.	
Scenario 2: Mid-block crosswalk + pedestrian	The driver is on a straight four-lane road, two lanes in either direction. There is a mid-block crosswalk ahead (downstream of the truck). There is also a pedestrian ahead sign that is on the right-hand side of the road. The participant must scan toward the right for the pedestrian ahead sign, and then scan to the right as the truck is passed for a potential pedestrian.	
Scenario 3: Midblock crosswalk	In this scenario, the participant is driving on a two-lane road with one travel lane in either direction. The truck in the opposing lane is waiting to take a right turn into the parking lot. The truck is stopped just after a crosswalk obscuring the view of a potential pedestrian on the left, toward which the driver should scan when passing the truck.	
Scenario 4: Left turn at four-way stop controlled intersection + vehicle in opposing left turn lane	The driver is taking a left turn at a signalized intersection. A large truck across the intersection in the left turn lane obscures a motorcyclist who might be passing the truck on its right-hand side, potentially colliding with the turning driver. The driver should slow and glance toward the left of the truck as he or she completes the turn to the left.	
Scenario 5: Signal controlled intersection + line of vehicles	There is a line of vehicles in the right lane of a roadway with two travel lanes in either direction. A signalized intersection is ahead. The driver is in the left lane. A vehicle ahead and in the right, may change lanes and move into the left lane immediately ahead of the driver. The participant needs to scan toward the right lane for vehicles that may emerge as a potential hazard.	
Scenario 6: Road entering from the left-hand side	The driver passes a traffic sign that shows a road entering from the left-hand side. Often such signs are present because drivers entering from the side road are difficult to see (obscured by vegetation or geometry) or are unexpected. The driver should glance to the left for potential vehicles entering from that direction.	

feedback driven (*5, 6*).

V-RAPT Scenarios. As noted above, a total of six scenarios were developed for V-RAPT. They are displayed below in Table 1. Descriptions are given of each scenario, the latent threat, and what the driver should do.

V-RAPT Training. As noted above, V-RAPT has four different modular phases for each of the six scenarios chosen for training. The first module is the mistake module. In this module, the participants navigate through each of the virtual scenarios using the Oculus Rift. Their drive in the first section is recorded for subsequent reference in the other modules. The participants can control the speed of the vehicle, but not the location.

In the second module, the participant is first trained about the latent hazard, training which is specific to each scenario. The instructions given in the training section for each of the six scenarios are provided in Table 2. These instructions are given while a video (not the participants) is played in real time from the driver's perspective of the scenario. The instructions include information about the feature of the environment which suggests that a latent threat appears ("approaching launch area", i.e., the area where the driver should "launch" an eye movement or head turn), information about where the driver should look and what the driver

Table 2. Audio Script in the Mediation Section

Scenario	Approaching launch area	Beginning of launch zone	End of launch zone
Scenario 1: Curve + T Intersection	There is a sign on the right indicating that pedestrians are crossing the road somewhere ahead of you. You should start scanning the forward roadway for a crosswalk.	You can just barely see a crosswalk ahead of you in front of the car in the opposing lane. However, your view of the pedestrian on the right-hand side of the road could be hidden by the truck on the right waiting to turn. You should keep scanning toward the front edge of the truck on the right for a pedestrian while keeping your speed slow.	Keep your speed slow and keep scanning toward the front edge of truck on the right for any potential pedestrian that might enter the crosswalk.
Scenario 2: Mid-block crosswalk + pedestrian	There is a sign on the right indicating that pedestrians are crossing the road somewhere ahead of you. You should look for an obvious place where that might occur.	The truck parked in the right most lane can block your view of a potential pedestrian entering the crosswalk in front of the truck. You should keep scanning toward the right front edge of the truck and you should slow down.	Keep your speed slow and continue scanning toward the truck on your right for a pedestrian that might enter the crosswalk.
Scenario 3: Midblock crosswalk	There is a sign on the right indicating that pedestrians are crossing the road ahead. You should keep scanning toward the left- and right-hand side of the road in the area of the crosswalks for pedestrians.	There is a sign on the right indicating a pedestrian crosswalk ahead. The truck in the opposing left lane waiting to make a turn toward the parking lot on your right is blocking your view of a potential pedestrian behind the truck who is in the crosswalk. You should slow down and scan toward the left most and right most edge of the truck for any obscured pedestrian.	You should keep scanning for any potential pedestrian by the truck while keeping your speed slow.
Scenario 4: Left turn at four-way stop controlled intersection + vehicle in opposing left turn lane	Because you are turning left, you want to glance at traffic across the intersection that might collide with your vehicle.	The truck in the opposing left lane might obstruct your view of other vehicles in the lane adjacent to the truck. These other vehicles could strike you as you are turning left. You should slow down and look to the right.	As you proceed to turn left, slow down enough until you can determine whether there is any oncoming traffic on the right hidden by the trucks.
Scenario 5: Signal controlled intersection + line of vehicles	The signal in the upcoming intersection is red. You should watch for vehicles that might change lanes in front of you as you approach the signal.	The vehicle in front of you has a clear path to through the intersection if the driver changes into your lane and may be in a hurry. You should continue to glance toward this vehicle for any possible sudden moves.	As you are passing the vehicle keep scanning toward your right as you might be in the blind spot of the vehicle on your right.
Scenario 6: Road entering from the left-hand side	There is a sign on the right-hand side indicating that traffic may be entering from the left. You should be alert at this point and keep scanning for where traffic might enter from the left.	The traffic entering from the left is obscured by trees. The trees might hide your view of the driver of the vehicle waiting to merge into your lane. Slow down and keep scanning on your left for any vehicle trying to merge into your lane.	Keep your speed slow and keep scanning toward the edge of the tree line on your left for entering vehicles.

should do in the immediate vicinity of the launch area, and information about the driver should do after exiting the launch area. Immediately after receiving information about the potential threat, the participants are shown the recorded video from the first module. If the participant correctly scanned for the latent hazard, he or was complimented for the good performance and safe driving. However, if the participant failed to make a correct anticipatory glance at the latent hazard, then a general description about the latent hazard in the current scenario was provided. Modules 2 and 3

Table 3. List of Simulator Scenarios with Perspective Views

Scenario name	Scenario description	Perspective scenario views
N1 (Training Scenario1)	The driver is approaching a T-intersection where there is a vehicle waiting in the forward lane in the opposite direction and another vehicle on the connecting road (cross street). The vehicle on the cross street is blocking the view of a potential pedestrian. The driver needs to scan for the pedestrian.	
N2 (Training Scenario 2)	The driver on a straight four-lane road, two lanes in either direction. There is a mid-block crosswalk ahead (downstream of the truck). There is also a pedestrian ahead sign that is on the right-hand side of the road. The participant must scan toward the right for the pedestrian ahead sign, and then scan to the right as the truck is passed for a potential pedestrian.	
N3 (Training Scenario 3)	In this scenario, the participant is driving on a two-lane road with one travel lane in either direction. The truck in the opposing lane is waiting to take a right turn into the parking lot. The truck is stopped just after a crosswalk obscuring the view of a potential pedestrian on the left, toward which the driver should scan when passing the truck.	
N4 (Training Scenario 4)	The driver is taking a left turn at a signalized intersection. A large truck across the intersection in the left turn lane obscures a motorcyclist who might be passing the truck on its right-hand side, potentially colliding with the turning driver. The driver should slow and glance toward the left of the truck as he or she completes the turn to the left.	
N5 (Training Scenario 5)	There is a line of vehicles in the right lane of a roadway with two travel lanes in either direction. A signalized intersection is ahead. The driver is in the left lane. A vehicle ahead and in the right may change lanes and move into the left lane immediately ahead of the driver. The participant needs to scan toward the right lane for vehicles that may emerge as a potential hazard.	
N6 (Training Scenario 6)	The driver passes a traffic sign that shows a road entering from the left-hand side. Often such signs are present because drivers entering from the side road are difficult to see (obscured by vegetation or geometry) or are unexpected. The driver should glance to the left for potential vehicles entering from that direction.	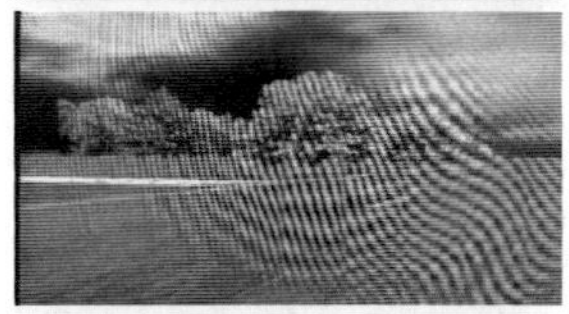
F1	The scenario begins with the participant driving down a four-lane road (two travel lanes in each direction). There is a parking lot on the right-hand side of the road. A car is waiting to pull out from the parking lot. The driver needs to pay attention to the right-hand side of the road for any vehicle that may pull out from the parking lot.	
F2	The scenario starts on a two-lane curved road. As the driver approaches the apex of the curve, immediately following the apex there is a truck in the emergency lane with its emergency flashers activated just in front of a crosswalk. The driver needs to pay attention to pedestrians that might emerge from in front of the truck.	
F3	The driver is on a two-lane suburban road. The driver passes a traffic sign that indicates that pedestrians may be present at the school zone. Ahead is a bus that is stopped on the left-hand side at a marked mid-block crosswalk for a potential pedestrian that may enter the crosswalk.	
F4	The driver is travelling on a four-lane roadway. A vehicle on the right at the intersection is obscured from the driver by another large vehicle on the right. As the driver will go straight through this intersection, the driver cannot see past the truck to where a car or motorcyclist might be passing the truck and might emerge as a potential hazard.	

combine to provide mediation to the user. The fourth module is a mastery section, in which the participant again navigated the driving scenario in the Oculus Rift. While navigating, if the participant made the correct anticipatory glances at the target zone, the participant was assigned the next training scenario, and if the participant failed to make the correct anticipatory glance then the participant was retrained on the same scenario. There is a total of six training scenarios administered in a modular manner. The scenarios are counterbalanced across subject and the four modules were delivered for each scenario individually.

Risk Awareness and Perception Training. RAPT was developed and tested at the University of Massachusetts Amherst and was used to study the effect of providing risk perception training to novice drivers. It is a PC-based program using static views that has been described in detail elsewhere (*32*).

Placebo Training. The placebo training program is also a PC-based training program and is designed and presented in a modular manner like the other training programs. In this program, the driver navigates through the virtual drives on the PC. However, unlike in the V-RAPT and RAPT training, placebo trained drivers receive no feedback on their performance and no instruction on how to anticipate and mitigate the latent hazard and become a safer driver. Further, the placebo training also presents information on how to gauge tire pressure. On average, the program takes 30 minutes to complete.

Simulator Evaluation Scenarios

Following training, to evaluate the effectiveness of the training programs, the participants were asked to navigate 10 micro-scenarios (six near transfer scenarios, N1 to N6, and four far transfer scenarios, F1 to F4) each of which lasted approximately two minutes in duration. All 10 scenarios contained a single latent hazard and were counterbalanced using a Latin square. The scenarios are briefly described along with their perspective views for illustration in Table 3. The descriptions of the first six simulator scenarios are identical to those of the first six training scenarios, although the virtual worlds are quite different in detail (compare the perspective views). The descriptions are included a second time to make it easier for the reader to map the training to the simulator scenarios.

Experimental Design

The experiment utilizes a between-subject design. The participant is either trained in the VRH-based hazard anticipation and mitigation training program (V-RAPT), the PC-based hazard anticipation (RAPT) training program, or PC-based placebo training program. After taking the appropriate training program, all the participants navigated 10 scenarios (six near transfer and four far transfer) on the driving simulator. The ordering of the 10 simulator scenarios was completely counterbalanced across participants using a Latin square (*32, 33*).

Procedure

Participants first completed an informed consent form to provide written consent to participate in this study. Then, they were randomly assigned to one of three training programs. After the completion of the training program, participants were fitted with an eye tracker, had their eyes calibrated, and were provided with a practice drive in the simulator to get them acclimated to the controls of the simulator cab and the simulator environment. Following this, all participants navigated 10 virtual scenarios on the simulator. They also completed a single post study questionnaire and a stipend voucher. The entire study session ranged in duration from an hour to an hour and a half, depending on the training group that the participants were assigned to.

Dependent Variables

Two dependent variables were measured, one indexing hazard anticipation and one indexing hazard mitigation. First, the dependent variable indexing latent hazard anticipation for each scenario was whether the driver glances toward the potential hazard. A driver's latent hazard detection performance for each scenario is manually scored as either a 0 (miss) if he or she fails to glance toward the target zone in the launch zone, or a 1 (hit) if he or she successfully glances toward the target zone whilst in the launch zone. Specifically, a target zone was defined as that area of the forward roadway where a potential or actual threat may be present. A launch zone was defined as that area of the roadway from which the driver should glance toward the target zone to be able to successfully detect and mitigate for all latent hazard types. The launch zone and target zone used in this experiment were previously utilized and validated in other experiments for the evaluation of latent hazard anticipation (*34*).

A drivers' hazard mitigation behavior was captured starting about 100 feet before a latent hazard and ending 50 feet after the latent hazard by examining average velocity, the standard deviation of velocity, and average absolute acceleration. The average absolute acceleration is defined as the average of the absolute value of the acceleration at each point that was measured 100 feet before the latent hazard and 50 feet after the latent hazard. Previous work has shown that these measures are predictive of crashes and/or the severity of crashes (*35, 36*).

Hypotheses

There are two main hypotheses in this study. The first hypothesis is motivated by the enhancements to the 3M training discussed above. The second hypothesis follows because V-RAPT does train hazard mitigation in addition to hazard anticipation, whereas RAPT focuses only on hazard anticipation training.

- **H1**: The V-RAPT trained drivers will anticipate a greater proportion of the latent hazards than both the RAPT and placebo trained drivers in both the near transfer and far transfer evaluation scenarios.
- **H2:** The V-RAPT trained drivers will also demonstrate better hazard mitigation ability—as measured by their average velocity, standard deviation of velocity, and average absolute acceleration near the latent threat—than the RAPT and placebo trained drivers.

Results

Below, first the results on hazard anticipation are presented and then the results on hazard mitigation.

Hazard Anticipation Results

The binary-coded, binomially distributed eye movement data were analyzed using a logistic regression model within the framework of generalized estimating equations. The model included participants as a random effect, scenarios as a within subject variable, and treatment (three training groups) as a between-subjects factor. A significant main effect of treatment was observed, Wald $\chi^2(3) = 19.21$; $p < .001$. The main effect was consistent with our hypothesis. The V-RAPT trained drivers anticipated a greater proportion of latent hazards across all scenarios as compared to the RAPT trained drivers (87.57% vs 60.50%) which is a statistically significant difference, Wald $\chi^2(3) = 6.68$; $p < .001$. The RAPT trained drivers were also found to anticipate a significantly greater proportion (60.5% vs 28.88%) of hazards as compared to their placebo trained peers, Wald $\chi^2(3) = 21.83$; $p = .001$. Also significant is the difference in the proportion of hazards anticipated by the V-RAPT group compared to the placebo trained group (87.57% to 28.88%), Wald $\chi^2(3) = 19.21$; $p < .001$.

The same statistical model was used to observe separate hazard anticipation effects for the six near and four far transfer scenarios.

Near Transfer Scenarios. As noted above, the near transfer scenarios are the six simulator evaluation scenarios that are similar conceptually to the training scenarios. A significant main effect of treatment was observed for near transfer scenarios, Wald $\chi^2(3) = 26.94$; $p < .001$. The difference in the percentage of hazards anticipated in near transfer scenarios between V-RAPT and RAPT trained drivers (91.67% vs 57.12%—a difference of 34.55 percentage points) was statistically significant, Wald $\chi^2(3) = 15.80$; $p < .003$. The RAPT trained drivers anticipated a significantly higher proportion of hazards in the near transfer scenarios (57.12% vs 27.87%) as compared to their placebo trained peers Wald $\chi^2(3) = 76.47$; $p < .001$. The difference in the proportion of hazards anticipated by the V-RAPT trained drivers in the near transfer scenarios compared to the placebo trained drivers is statistically significant, being 63.8 percentage points, Wald $\chi^2(3) = 106.18$; $p < .001$.

Far Transfer Scenarios. Far transfer scenarios include the four simulator evaluation scenarios that were not closely related to the training scenarios. A significant main effect of treatment was also observed for far transfer scenarios using the same logistic regression model, Wald $\chi^2(3) = 26.34$; $p < .001$. The V-RAPT trained drivers anticipated a greater proportion of latent hazards across the four far transfer scenarios (82.50%) compared to the RAPT trained drivers (65.15%), Wald $\chi^2(3) = 10.24$; $p = .009$. Both the RAPT (Wald $\chi^2(3) = 13.21$; $p < .005$) and V-RAPT trained (Wald $\chi^2(3) = 74.69$; $p < .001$) drivers anticipated a significantly greater proportion of latent hazards across all four far transfer scenarios compared to their placebo trained peers (30.3%).

Hazard Mitigation Results

The average velocity between a point about 100 feet before a latent hazard and a point 50 feet after the latent hazard was analyzed using analysis of variance (ANOVA) with treatment (training) as a between-subjects factor. A main effect of treatment was revealed, $F(2, 316) = 9.94$, $\eta^2 = 0.99$, $p < .005$. The average velocities of the V-RAPT ($M = 32.42$, $SD = 7.39$) and placebo ($M = 37.69$, $SD = 4.82$) groups, $F(1, 210) = 20.60$, $\eta^2 = 0.994$, $p < .005$, did differ significantly, suggesting V-RAPT trained drivers learn to mitigate hazards by driving slowly when approaching the hazard. However, the average velocities of V-RAPT and RAPT ($M = 34.18$, $SD = 6.49$) trained drivers did not differ significantly [$F(1, 209) = 2.07$, $\eta^2 = 0.951$, $p = 0.15113$], suggesting that the RAPT and V-RAPT trained drivers are equally good at hazard mitigation. Finally, the average velocities of RAPT and placebo trained drivers were looked at and the differences were significant, $F(1, 213) = 8.29$, $\eta^2 = 0.988$, $p < .005$, suggesting RAPT trained drivers mitigate hazard significantly better compared to placebo trained drivers'.

The standard deviation of velocity between a point 100 feet before a latent hazard and a point 50 feet after the latent hazard was analyzed with the same ANOVA model. Again, a main effect of treatment was revealed, $F(2, 316) = 9.22$, $\eta^2 = 0.979$, $p < .005$. The standard deviations of velocity of the V-RAPT ($M = 7.39$, $SD = 3.94$) and placebo ($M = 4.82$, $SD = 4.48$) groups, $F(1, 210) = 19.51$, $\eta^2 = 0.974$, $p < .005$, did differ significantly. Additionally, the standard deviations of velocity of the V-RAPT and RAPT ($M = 6.49$, $SD = 4.78$) trained drivers did not differ significantly [$F(1, 209) = 2.18$, $\eta^2 = 0.678$, $p = 0.140$], suggesting the RAPT and V-RAPT trained drivers are no different at mitigating hazards with regard to their modulation of velocity near the latent threat.

Finally, the standard deviations of velocity of the RAPT and placebo trained drivers were looked at and the differences were significant, $F(1, 213) = 6.98$, $\eta^2 = 0.921$, $p < .005$, suggesting RAPT trained drivers mitigate hazards significantly better than placebo trained drivers.

Finally, the average absolute acceleration between a point 100 feet before a latent hazard and a point 50 feet after the latent hazard was analyzed with the same ANOVA model with treatment as a between-subjects factor. Again, a main effect of treatment was revealed, $F(2, 316) = 10.58$, $\eta^2 = 0.987$, $p < .005$. The average absolute accelerations of the V-RAPT ($M = 0.64$, $SD = 0.53$) and the placebo ($M = 0.42$, $SD = 0.35$) groups, $F(1, 210) = 22.71$, $\eta^2 = 0.973$, $p < .005$, differed significantly. Additionally, the average absolute accelerations of the V-RAPT and RAPT ($M = 0.54$, $SD = 0.46$) trained drivers differed significantly ($F(1, 209) = 4.32$, $\eta^2 = 0.727$, $p = 0.038$. Finally, the average absolute accelerations of the RAPT and placebo trained drivers were significantly different $F(1, 213) = 6.00$, $\eta^2 = 0.946$, $p < .015$.

Discussion

The current experiment investigates the effectiveness of the newly developed, VRH-based hazard anticipation and hazard mitigation training program (V-RAPT) for young drivers. Previous studies have shown that the young driver fails to scan adequately for latent hazards (*5*). Moreover, it has been shown that young drivers can be trained to double the likelihood that they scan for latent hazards, reducing the gap between untrained novice drivers and experienced drivers by half in just an hour of training (*9*). However, this still left lots of room for improvement. This was why V-RAPT was developed, in theory enhancing the mediation that is delivered and thereby the value of training. Consistent with the first hypothesis, drivers that received V-RAPT anticipated a significantly greater proportion of latent hazards compared to the placebo trained driver and the RAPT trained drivers. In particular, V-RAPT almost tripled the performance of the untrained novice drivers, considerably higher than is typically observed in the evaluation of similar hazard anticipation training programs delivered on other platforms (*5, 9, 12*). Further, the results demonstrate that participants trained on V-RAPT anticipate a greater proportion of latent hazards both on scenarios which are similar (near transfer) to those trained on, and on scenarios dissimilar (far transfer) from those trained on. Transferability is an important characteristic to assess the effectiveness of the training because, ultimately, there are only a finite number of situations that can be trained on and evaluated for in a controlled manner. The proportion of latent hazards anticipated by the RAPT trained drivers was 60.50% and was in line with that shown by previous studies (*4, 22, 29*).

To examine the second hypothesis, three related measures of driver vehicle behaviors were analyzed. All were consistent with the superiority of V-RAPT to no training. Surprisingly, the improvement in the hazard mitigation behavior of the V-RAPT trained drivers did not differ from that of the RAPT trained drivers when either speed or the standard deviation of speed was used as the dependent variable, but did differ when the absolute acceleration was used as the dependent variable. Three points are worth discussing. First, although the differences between the V-RAPT and RAPT groups when the dependent measures were speed and the standard deviation of speed were not statistically significant, the direction of the differences was as predicted.

Second, no previous studies had evaluated the effect of hazard anticipation training alone on hazard mitigation behaviors. Thus, as participants could learn both information about how better to anticipate hazards and mitigate those hazards (V-RAPT) in the same time as they were able to learn only about hazard anticipation (RAPT) indicates that V-RAPT does not increase hazard anticipation skills at the expense of hazard mitigation skills.

Third, the finding that the absolute acceleration differentiates the V-RAPT trained drivers from the RAPT trained and placebo trained drivers is worth a brief comment, even if it is only speculative at this point. Drivers who slow less will have a smaller standard deviation of velocity. This would explain why the RAPT drivers have a smaller standard deviation of velocity than the V-RAPT trained drivers. Moreover, if the drivers who are in the V-RAPT condition slow gradually whereas the drivers who are in the RAPT conditions slow precipitously in the presence of the latent threat, then the absolute acceleration will be larger for drivers in the V-RAPT condition than for drivers in the RAPT condition.

In summary, this study shows that a VRH-based hazard anticipation and hazard mitigation training program can lead to potentially much larger improvements in these behaviors than training programs delivered on other platforms drivers (*4, 20, 27, 28*). Additionally, it is important to comment on simulator sickness because this could be a barrier to adoption of programs like V-RAPT. Typically, the reported simulator sickness rates in virtual reality, headset-based interventions are very high. But, in this experiment, only a single V-RAPT trained participant dropped out because of simulator sickness. There may be several reasons for the observation of low simulator sickness rates including the use of optimized micro-scenarios (scenarios which occurred over seconds instead of minutes or hours), and the provision of short 30 second breaks between each scenario, or if required, between the different modules of the scenario, a proven method for reducing simulator sickness (*37*). Another reason for the low simulator sickness rates of the V-RAPT group could be the specific instructions provided to participants to not make sudden and jerking head movements during the training simulation. Also, the participants in this study were young novice drivers, and as literature states the older drivers are more proven to simulator sickness as compared to the young novice drivers

Limitations

There are important limitations associated with the training program. The V-RAPT training program currently lacks a user interface which is entirely automated. The interface as it is configured now requires an instructor to always be present. Having said this, although the instructor needs to start and stop the scenarios, the training instructions provided during the scenarios are incorporated into the virtual scenarios, in the form of computer-readable audio files.

The current evaluation examines the effectiveness of training for young drivers (18–25 years old). However, it is novice drivers in their teens who are at most risk. Also, the number of scenarios used in training and in the near and far evaluation of the effectiveness is relatively small and not necessarily representative of the types of crashes in which young drivers are overrepresented. The number of teens is small and certainly not representative of the entire population of drivers. There was no assessment of the long-term retention of the training. Finally, there was no assessment in the field of the effect of training on hazard anticipation and hazard mitigation training or of the effect on crashes.

Acknowledgments

The current research was supported by grants from State Farm and the US Department of Transportation Tier 1 SaferSim University Transportation Center (UTC).

References

1. Centers for Disease Control and Prevention. *Teen Drivers: Get the Facts*, 2015. https://www.cdc.gov/Motorvehiclesafety/teen_drivers/teendrivers_factsheet.html
2. National Highway Traffic Safety Administration. *Research Agenda for an Improved Novice Driver Education Program.* Report to Congress DOT-HS-808-161, Department of Transportation, National Highway Traffic Safety Administration, Washington, D.C., 1994.
3. McKnight, J. A., and S. A. McKnight. Young Novice Drivers: Careless or Clueless. *Accident Analysis and Prevention*, Vol. 35, No. 6, 2003, pp. 921–925.
4. Crundall, D., and A. Pradhan. Hazard Avoidance: Definitions and a Framework for Research in Novice Drivers' Skills in Dealing with Roadway Hazards. In *Handbook of Teen and Novice Drivers* (D. Fisher, J. Caird, W. Horrey, and L. Trick, eds.), CRC Press, Boca Raton, 2016, pp 61–74.
5. Pradhan, A. K., K. R. Hammel, R. DeRamus, D. A. Noyce, A. Pollatsek, and D. L. Fisher. The Use of Eye Movements to Evaluate the Effects of Driver Age on Risk Perception in an Advanced Driving Simulator. Presented at 82nd Annual Meeting of the Transportation Research Board, Washington, D.C., 2003.
6. Fisher, D. L., N. E. Laurie, R. Glaser, K. Connerney, A. Pollatsek, and S. A. Duffy. The Use of an Advanced Driving Simulator to Evaluate the Effects of Training and Experience on Drivers' Behavior in Risky Traffic Scenarios. *Human Factors*, Vol. 44, 2002, pp. 287–302.
7. Regan, M. A., T. J. Triggs, and S. T. Godley. Simulator-Based Evaluation of the DriveSmart Novice Driver CD-ROM Training Product. *Proceedings of the Road Safety Research, Policing and Education Conference*, Brisbane, Queensland, Australia, 2000, pp. 315–320.
8. McKenna, F. P., and J. L. Crick. *Hazard Perception in Drivers: A Methodology for Testing and Training.* TRL Contractor Report (CR313), Transport Research Laboratory, Wokingham, 1994.
9. Taylor, T., K. Masserang, A. Pradhan, G. Divekar, S. Samuel, J. Muttart, A. Pollatsek, and D. Fisher. Long-Term Effects of Hazard Anticipation Training on Novice Drivers Measured on the Open Road. *Proc., 6th International Driving Symposium on Human Factors in Driving Assessment, Training, and Vehicle Design*, Lake Tahoe, CA, 2011, pp. 187–194.
10. Divekar, G., S. Samuel, A. Pollatsek, F. D. Thomas, K. Korbelak, R. D. Blomberg, and D. L. Fisher. On-Road Evaluation of a Hazard Perception Training Program Using Eye Tracking and Commentary Driving. Presented at 95th Annual Meeting Transportation Research Board, Washington, D.C., 2016.
11. Yamani, Y., P. Bıçaksız, D. B. Palmer, J. Cronauer, and S. Samuel. Following Expert's Eyes: Evaluation of The Effectiveness of a Gaze-Based Training Intervention on Young Drivers' Latent Hazard Anticipation Skills. *Proc., 9th International Driving Symposium on Human Factors in Driver Assessment, Training, and Vehicle Design*, Manchester, VT, June 2017.
12. Zafian, T. M., S. Samuel, J. Coppola, E. G. O'Neill, M. R. Romoser, and D. L. Fisher. On-Road Effectiveness of a Tablet-Based Teen Driver Training Intervention. *Proceedings of the Human Factors and Ergonomics Society Annual Meeting*, SAGE Publications, September 2016, pp. 1926–1930.
13. Yamani, Y., S. Samuel, M. A. Knodler, and D. L. Fisher. Evaluation of the Effectiveness of a Multi-Skill Program for Training Younger Drivers on Higher Cognitive Skills. *Applied Ergonomics*, Vol. 52, 2016, pp. 135–141.
14. Roenker, D. L., G. M. Cissell, K. K. Ball, V. G. Wadley, and J. D. Edwards. Speed-of-Processing and Driving Simulator Training Result in Improved Driving Performance. *Human Factors*, Vol. 45, No. 2, 2003, pp. 218–233.
15. Pollatsek, A., D. L. Fisher, and A. K. Pradhan. Identifying and Remediating Failures of Selective Attention in Younger Drivers. *Current Directions in Psychological Science*, Vol. 15, 2006, pp. 255–259.
16. Zhang, T., J. Li, H. Thai, T. Zafian, S. Samuel, and D. L. Fisher. Evaluation of the Effect of a Novice Driver Training Program on Citations and Crashes. *Proceedings of the Human Factors and Ergonomics Society Annual Meeting*, SAGE Publications, September 2016, pp. 1991–1995.
17. Thomas, F. D., K. T. Korbelak, G. U. Divekar, R. D. Blomberg, M. R. E. Romoser, and D. L. Fisher. *Evaluation of a New Approach to Training Hazard Anticipation Skills of Young Drivers.* National Highway Traffic Safety Administration, Washington, D.C., 2013.
18. Thomas, F. D., S. Rilea, R. D. Blomberg, R. C. Peck, and K. T. Korbelak. *Evaluation of the Safety Benefits of the Risk*

Awareness and Perception Training Program for Novice Teen Drivers. Report No. DOT HS 812 235, U.S. Department of Transportation, Washington, D.C., 2016.

19. Vlakveld, W., M. R. E. Romoser, H. Mehranian, F. Diete, and D. L. Fisher. Does the Experience of Crashes and Near Crashes in a Simulator-Based Training Program Enhance Novice Driver's Visual Search for Latent Hazards? *Transportation Research Record: Journal of the Transportation Research Board*, 2011. 2265: 153–160.
20. Pradhan, A. K., A. Pollatsek, M. Knodler, and D. L. Fisher. Can Younger Drivers Be Trained to Scan for Information Which Will Reduce Their Risk Roadway Traffic Scenarios That Are Hard to Identify as Hazardous? *Ergonomics*, Vol. 52, No. 6, 2009, pp. 657–673.
21. Willis, D. K. The Impetus for the Development of a New Risk Management Training Program for Teen Drivers. *Proceedings of the Human Factors and Ergonomics Society 42nd Annual Meeting*, Human Factors and Ergonomics Society, Santa Monica, CA, 1998, pp. 1394–1395.
22. Fisher, D. L., J. Young, L. Zhang, M. Knodler, and S. Samuel. *Accelerating Teen Driver Learning: Anywhere, Anytime Training*. AAA FTS, Washington, D.C., 2017.
23. Young, J., T. Zhang, S. Samuel, M. Knodler, and D. L. Fisher. Simulator Evaluation of the Effectiveness of an Accelerated Teen Driver Training Intervention. Presented at 96th Annual Meeting of the Transportation Research Board, Washington, D.C., 2017.
24. Frese, M., and Altmann, A. The Treatment of Errors in Learning and Training. In *Developing Skills with New Technology* (L. Bainbridge and S. Quintanilla, eds.), Wiley, Chichester, 1989, pp. 65–86.
25. Ivancic, K., and B. Hesketh. Learning from Errors in a Driving Simulation: Effects on Driving Skill and Self-Confidence. *Ergonomics*, Vol. 43, No. 12, 2000, pp. 1966–1984.
26. Kulhavy, R. W. Feedback in Written Instruction. *Review of Educational Research*, Vol. 47, 1997, pp. 211–232.
27. Anguera, J. A., J. Boccanfuso, J. L. Rintoul, O. Al-Hashimi, F. Faraji, J. Janowich, E. Kong, Y. Larraburo, C. Rolle, E. Johnston, and A. Gazzaley. Video Game Training Enhances Cognitive Control in Older Adults. *Nature*, Vol. 501, No. 7465, 2013, pp. 97–101.
28. Rizzo, A. A., J. G. Buckwalter, T. Bowerly, C. Van Der Zaag, L. Humphrey, U. Neumann, C. Chua, C. Kyriakakis, A. Van Rooyen, and D. Sisemore. The Virtual Classroom: A Virtual Reality Environment for the Assessment and Rehabilitation of Attention Deficits. *CyberPsychology & Behavior*, Vol. 3, No. 3, 2000, pp. 483–499.
29. Lengenfelder, J., M. T. Schultheis, T. Al-Shihabi, R. Mirant, and J. DeLuca. Divided Attention and Driving: A Pilot Study Using Virtual Reality Technology. *The Journal of Head Trauma Rehabilitation*, Vol. 17, No. 1, 2002, pp. 26–37.
30. Thomson, J. A., A. K. Tolmie, H. C. Foot, K. M. Whelan, P. Sarvary, and S. Morrison. Influence of Virtual Reality Training on the Roadside Crossing Judgments of Child Pedestrians. *Journal of Experimental Psychology: Applied*, Vol. 11, No. 3, 2005, pp. 175–186.
31. McComas, J., M. MacKay, and J. Pivik. Effectiveness of Virtual Reality for Teaching Pedestrian Safety. *CyberPsychology & Behavior*, Vol. 5, No. 3, 2002, pp. 185–190.
32. Pradhan, A., D. Fisher, and A. Pollatsek. Risk Perception Training for Novice Drivers: Evaluating Duration of Effects of Training on a Driving Simulator. *Transportation Research Record: Journal of the Transportation Research Board*, 2006. 1969: 58–64.
33. Bradley, J. V. Complete Counterbalancing of Immediate Sequential Effects in a Latin Square Design. *Journal of the American Statistical Association*, Vol. 53, No. 282, 1958, pp. 525–528.
34. Samuel, S., and D. L. Fisher. Evaluation of the Minimum Forward Roadway Glance Duration. *Transportation Research Record: Journal of the Transportation Research Board*, 2015. 2518: 9–17.
35. Kockelman, K. K., and J. Ma. Freeway Speeds and Speed Variations Preceding Crashes, within and across Lanes. *Journal of the Transportation Research Forum*, Vol. 46, 2010, pp. 43–62.
36. af Wåhlberg, A. E. Driver Celebration Behavior and the Prediction of Traffic Accidents. *International Journal of Occupational Safety and Ergonomics*, Vol. 12, No. 3, 2006, pp. 281–296.
37. Schneider, C., F. S. Hajiseyedjavadi, J. Zhang, M. R. E. Romoser, S. Samuel, M. Knodler, and D. Fisher. Use of Micro Scenarios to Reduce the Effects of Simulator Sickness in Training Intervention Studies. *Proceedings of the Human Factors and Ergonomics Society Annual Meeting*, Washington, D.C., September 2016, pp. 1946–1950.

The Standing Committee on Operator Education and Regulation (ANB30) peer-reviewed this paper (18-01500).

The opinions, findings, and conclusions expressed in this publication are those of the authors and not those of the Department of Transportation or the Volpe National Transportation Systems Center.

Article

TRR
JOURNAL OF THE TRANSPORTATION RESEARCH BOARD

Transportation Research Record
2018, Vol. 2672(33) 31–41

DOI: 10.1177/0361198118789952
journals.sagepub.com/home/trr
SAGE

Safety of Freeway Service Patrol Operations: Case Study of Florida Road Rangers

Grady Carrick[1], Sivaramakrishnan Srinivasan[2], and Khajonsak Jermprapai[2]

Abstract

Safety service patrols are a proven strategy to mitigate the effects of traffic incidents through quick clearance, incident management, and assistance to other incident responders like police, fire, emergency medical services, and towing. As encountered by other responders, working on or near roadways presents unique hazards for safety service patrol vehicles and operators. Road Rangers are Florida's branded safety service patrols and, as a mature program with over 100 beats, a suitable case study for safety. This research combined an analysis of Road Ranger traffic crash data for 3 years with a comprehensive safety survey of more than 200 operators to determine safety characteristics related to service patrols. Comparing 200 Road Ranger traffic crashes from 2014 through 2016 with all Florida freeway crashes for the same time period revealed that Road Ranger crashes are five times more likely to involve a parked vehicle, and involve two or more vehicles 95% of the time. Pedestrian involvement, nighttime, shoulder locations, and work zones have higher representation for Road Ranger crashes, but weather is not a factor. Alcohol is three times more likely, drug use five times more likely, and distraction slightly higher when Road Ranger vehicles are struck. A survey of 217 Road Ranger drivers revealed that they are keenly aware of important safety topics like high-visibility safety apparel, non-traffic side vehicle approaches, and the dangers of working where there is limited lateral buffer space. Drivers overwhelmingly believe that they have the training and equipment necessary to do their jobs safely.

The Florida Department of Transportation (FDOT) contracts safety service patrols under the well-known branding "Road Rangers." Road Rangers assist motorists with vehicle disablements, remove debris from the roadway, and actively support incident responders with temporary traffic control and incident management at collisions. Florida's Road Rangers patrol more than 1,500 centerline freeway miles with approximately 105 beats covering a variety of days of the week and hours of the day. During 2013, Road Rangers made 374,971 assists and since the program inception they have made over 4.3 million assists (*1*). A 2010 study funded by the FDOT found the benefit–cost ratio of the entire Road Ranger program to be 6.68:1 (*2*). According to surveys of police, fire, transportation, and towing responders in Florida, more than 93% think that Road Rangers improve safety at incident scenes, and nine out of 10 responders believe Road Rangers help them do their job better (*3*).

There are significant hazards associated with working near moving traffic. Whether on an urban freeway or a rural local street, traffic incident responders are among the most vulnerable road users, challenged with working in dynamic and dangerous conditions that are neither designed nor intended for pedestrians. According to Florida traffic crash statistics, during 2016 there were 76 traffic crashes involving incident responders who were struck by vehicles.

Although there is no national reporting system for incident responders who are killed while working at the roadside, some insight can be obtained from industry sources. According to the National Law Enforcement Memorial Foundation, 134 police officers were killed in struck by incidents between 2005 and 2014, or about 13 per year (*4*). The National Fire Protection Association completed a study of fire personnel who were struck and killed by vehicles in 2014 and found that in the most recent 10 years, a total of 28 personnel were struck and killed by non-fire vehicles, or about three each year (*5*).

[1]Enforcement Engineering, Inc., Saint Johns, FL
[2]University of Florida, Gainesville, FL

Corresponding Author:
Address correspondence to Grady Carrick: gcarrick@enfenginc.com

The towing profession loses "about 40 to 50" operators per year, and it is estimated that about ¾ of those deaths involve struck by incidents (*6*).

The number of transportation workers killed on the job is difficult to pin down because in addition to government employees, there are many private industry contractors involved. In the work zone setting, more than 100 workers lose their life annually, with nearly one quarter of those being struck or run over by non-construction vehicles (*7*).

There is no published number of deaths involving safety service patrol operators. In conjunction with this project, a survey of national points of contact for safety service patrols, and a search of media accounts revealed that at least 20 service patrol operators have been killed in the line of duty since the inception of service patrols. These include four in Florida, three in Illinois, two in each of Georgia, Missouri, and Tennessee, and one in each of California, Louisiana, New Jersey, North Carolina, Texas, Virginia, and West Virginia.

Safety Service Patrols (SSP) have proven to be an effective means by which traffic incidents can be managed and congestion mitigated, saving lives and money. The job of SSP is dangerous, and it is important to gain insight into the things that present risk for SSP operators. This research has the objective to understand the characteristics of Florida Road Ranger crashes and related safety concerns from the perspective of operators, to help identify countermeasures and thereby improve safety.

Literature Review

SSP have been in existence since the 1960s, though they have really gained in popularity just in the past two decades (*8*). Though there are several government publications that guide implementation of SSP, there are no published studies about safety service patrol crash experience. There are however, related studies for other types of traffic incident responders that establish examples of how safety might be evaluated. It is useful to note that this study does not focus on the impact of service patrols on the overall safety of freeway operations and so the corresponding literature is not reviewed here.

Carrick, et al. examined more than 31,000 law enforcement vehicle collisions occurring in Florida from 2005 through 2008 (*9*). The study found that most law enforcement crashes are low-speed events on local streets. The law enforcement vehicles commonly were often stopped or parked when they crashed. Rear end crashes were common, but on freeway shoulders not prominent. Most crashes occurred during the daytime, under favorable driving conditions, and during routine operating mode. The environmental conditions for emergency mode were not significantly different than for other crashes.

The U.S. Fire Administration studied emergency vehicle markings, designed to increase visibility and promote safety of responders at highway incidents (*10*). Though the study points out that there is no national standard for marking responder vehicles, marking vehicles is a passive treatment that can ultimately prevent them from being struck while stationary at an incident scene. Making emergency vehicles conspicuous involves using contrasting colors, fluorescent for daytime, and retro-reflective for nighttime. The study encouraged the use of conspicuity markings, focused primarily on the rear of the vehicle.

Grant and Merrifield performed research for the Fire Protection Research Foundation, examining the sources of traffic crash report data related to ambulances (*11*). Ambulance traffic crash data were noted as fairly robust at the national and state level, though not a lot of analysis has been done. Model Minimum Uniform Crash Criteria (MMUCC) contain guidance for state crash reporting systems related to vehicle types and purposes that make finding ambulance crashes fairly easy (*12*).

Maistros and Schneider examined fatal law enforcement vehicle traffic crashes using Fatality Analysis and Reporting System (FARS) data for a 3-year period (*13*). A regression model found several factors to be significant, among which are the use of seat belts by officers, the presence of more than one person in the vehicle, and speed.

Savolainen et al. examined emergency vehicle crashes in Michigan using state traffic crash report records (*14*). The research found that about 28% of crashes involved emergency operation. Intersections were well represented and severity was greatly reduced when speeds were lower. The need to further differentiate police, fire, and ambulance vehicles was noted.

Methodology

Because research evaluating SSP safety is scarce, the approach here relies on official traffic crash reports and the views of individual operators in order to better understand safety. Examining historical traffic crashes can illuminate the characteristics of those events involving SSP operators. Surveying individual Road Ranger drivers can provide insight from their "boots on the ground" experience that comes from working around moving traffic every day. Correspondingly, this study used both crash data and a survey of Road Rangers to understand the safety of the service patrol operations. With an understanding of safety through this research, countermeasures can be recommended for improvement.

Analysis of Florida Traffic Crash Data

The research team also obtained traffic crash report information for incidents involving a Road Ranger service patrol (RRSP). During the course of normal operations, RRSP vehicles are sometimes involved in traffic collisions, most notably when they are struck while stopped at an incident scene. In rare cases, the RRSP operator may be involved in a vehicular collision while a pedestrian on or near the roadway, and these cases are also captured as a traffic crash. With the date, time, and location of a traffic crash, the research team can retrieve crash information utilizing the Signal 4 Analytics database which is housed at the University of Florida (*15*). Signal 4 Analytics is a repository of more than 5 million historical Florida Traffic Crash Report forms. Each crash record contains the more than 100 data elements from the individual crash report as well as an image of the actual form.

Individual FDOT districts and their RRSP contractors were polled to identify crashes involving RRSP for calendar years 2014–2016. FDOT's Sunguide Traffic Management Center (TMC) software contained a record of most incidents and RRSP contractors verified and supplemented the TMC list. Given that they maintain a complete accounting of crashes involving their vehicles and personnel for insurance purposes, the combination of these sources accounted for all RRSP crashes. For the 3-year period, 2014–2016, there were 200 RRSP crashes in Florida, resulting in 107 injuries and two fatalities. This represents an average of just over one RRSP crash every 5 days.

Most traffic crash report data is presented as nominal/categorical data, readily lending itself to presentation using descriptive techniques such as frequency distributions. Because RRSP operate only on freeways, a comparison with other freeway traffic collisions seems like a logical approach. For the same 3-year period, there were 176,993 crashes on Florida freeways. Comparing the frequencies for RRSP and all freeway crashes has the potential to show how they differ, though tests for statistical strength of the comparison are not appropriate given the relatively small number of RRSP crashes. Clear distinctions are readily apparent when comparing frequencies for the two sets.

Time of day and day of the week are fundamental traffic crash attributes. The day of the week is fairly consistent between the two sets of data with the noticeable exception of Monday, for which fewer RRSP crashes occur, and Wednesday, on which just over one in five RRSP crashes happens. There does not appear to be any explanation for these differences.

In Figure 1, one can see that the time of day differences show that early morning hours tend to be more dangerous for RRSP, as compared with the larger data set. The highest number of RRSP crashes were observed at 5 a.m., 1 p.m., 3 p.m., 8 p.m., and 10 p.m., perhaps coinciding with greater exposure associated with RRSP shift changes. Shifts throughout the state typically start at 5 or 6 a.m. and end at some time between 6 and 10 p.m., with a mid-day shift change. There are locations with 24-h coverage, as reflected in the data.

Crash Attributes

The type of crash data element is a variable to explain the manner of collision, typically between vehicles. Like most freeway incidents, the rear end crash is very

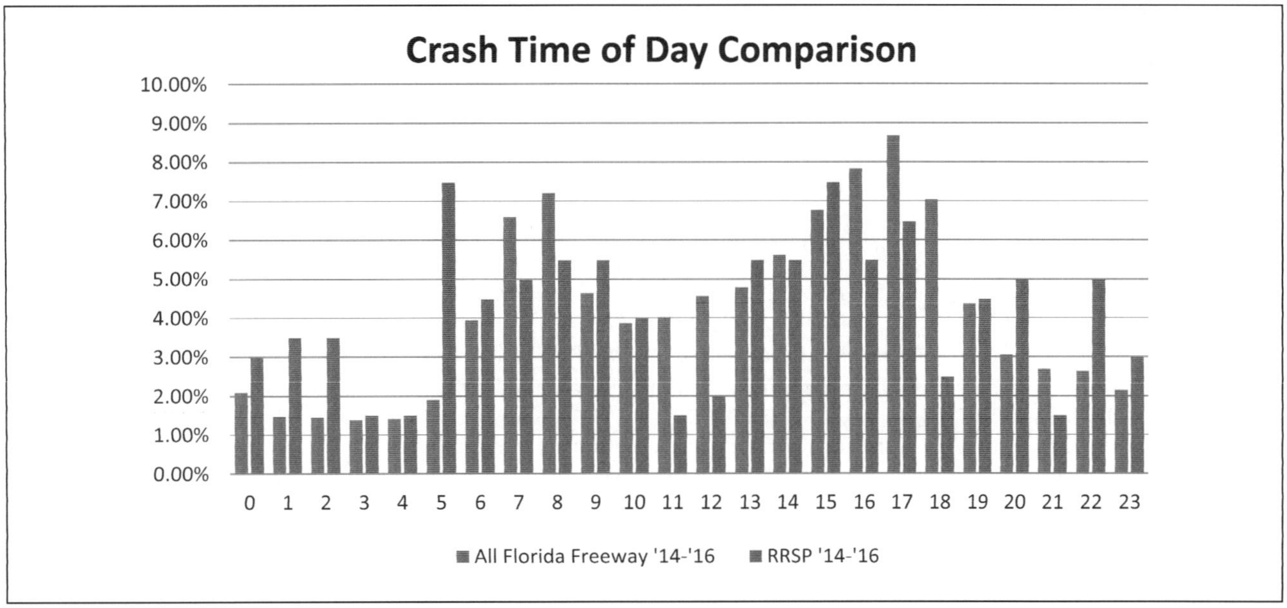

Figure 1. Time of day comparison between RRSP crashes and all freeway crashes in Florida.

Table 1. Event Characteristics Compared

Data element	Attribute	Florida freeway crashes 2014–2016 n = 176,993	RRSP crashes 2014–2016 n = 200
Crash_type	Angle	0.38%	2.49%
	Head on	0.26%	0.50%
	Left turn	0.30%	1.00%
	Off road	16.31%	4.48%
	Other	11.56%	25.37%
	Pedestrian	0.18%	3.48%
	Rear end	48.98%	43.28%
	Right turn	0.07%	0.50%
	Rollover	2.17%	1.00%
	Sideswipe	17.24%	15.92%
	Unknown	2.08%	1.99%
Vehicles	1	24.20%	5.47%
	2	63.52%	77.61%
	3 +	12.28%	16.91%
Non_motorists	0	99.71%	93.53%
	1	0.26%	5.47%
	2	0.02%	1.00%
Crash_type_detailed	Backed into	0.29%	1.49%
	Head on	0.26%	0.50%
	Left entering	0.21%	1.00%
	Off road	16.31%	4.48%
	Opposing sideswipe	0.27%	0.50%
	Other	3.93%	2.99%
	Parked vehicle	0.67%	19.40%
	Pedestrian	0.18%	3.48%
	Rear end	48.98%	43.28%
	Right angle	0.38%	2.49%
	Right/through	0.05%	0.50%
	Rollover	2.17%	1.00%
	Same direction sideswipe	16.97%	15.42%
	Single vehicle	6.67%	1.49%
Crash_severity	Fatality	0.52%	0.50%
	Injury	26.45%	37.81%
	Property damage only	73.03%	61.69%
Manner_of_collision	Angle	5.12%	7.46%
	Front to front	0.43%	1.00%
	Front to rear	47.19%	51.24%
	Other	28.82%	17.41%
	Rear to rear	0.06%	1.49%
	Sideswipe, opposite direction	0.14%	0.50%
	Sideswipe, same direction	16.74%	19.90%
	Unknown	1.29%	1.00%

common for RRSP. Run off road crashes are three times more likely among generic freeway crashes, and "other" types are twice as likely for RRSP crashes, probably explained by the varied duties of the job. Because the RRSP operators are almost always out of their vehicle at some point during their duties, pedestrian crashes involving RRSP account for 3.8% of crashes whereas in the typical freeway incident they are quite rare (0.18%). This is even more pronounced when "non-motorists" are counted in 6.47% of RRSP crashes and only 0.28% of the statewide freeway crashes.

The detailed crash type is equally illuminating. Again, run off road and single vehicle crashes are much higher in the statewide freeway crash set. Rear end crashes occur at a similar rate, but nearly one in five RRSP crashes involves a parked vehicle whereas they are quite rare (0.67%) in the larger data set. Same direction sideswipe and rear end types are fairly similar.

Whereas one in every four freeway crashes is a single vehicle event, more than 95% of RRSP crashes involve two vehicles or more vehicles. Road Ranger crashes involve three or more vehicles at a slightly higher frequency. Table 1 represents selected data elements that describe the collision event. Note that variable attributes are truncated so small frequency values are not listed, and consequently percentages do not always total 100%.

Table 2. Environmental Characteristics Compared

Data element	Attribute	Florida freeway crashes 2014–2016 n = 176,993	RRSP crashes 2014–2016 n = 200
Weather_condition	Clear	60.10%	66.67%
	Cloudy	23.76%	19.90%
	Rain	15.70%	12.94%
Light_condition	Dark: lighted	18.01%	27.86%
	Dark: not lighted	6.78%	4.48%
	Dawn	2.23%	1.99%
	Daylight	69.05%	59.70%
	Dusk	3.75%	5.47%
Road_surf_cond	Dry	75.26%	79.10%
	Wet	24.56%	20.40%
Contrib_circum_road1	Debris	1.18%	0.50%
	None	89.87%	90.55%
	Obstruction in roadway	0.58%	1.49%
	Road surface condition	6.60%	5.47%
	Work zone	1.29%	1.49%
Contrib_circum_env1	None	91.65%	94.03%
	Weather conditions	7.33%	5.47%
Work_zone_related	N	93.91%	89.55%
	Y	6.07%	9.95%

Environment

Weather plays a less significant role in the RRSP crash. Whereas rain is noted in 15.7% of freeway crashes, it is only present in 12.9% of RRSP crashes. Similarly, wet roads account for more than one in four freeway crashes (24.5%), but only 20.4% of those involving RRSP. Weather contributes to 7.3% of freeway crashes versus only 5.47% of RRSP events.

Lighting conditions are another common explanatory crash characteristic. Statewide freeway crashes are more likely to occur during daylight hours (69.1% versus 59.7%), but RRSP crashes are more likely to occur on dark roadways with artificial lighting (27.9% versus 18.0%). Periods of dusk and dawn are represented slightly more highly in RRSP crashes.

Road Ranger crashes are more likely to occur in or near a work zone (10.0% versus 6.1%) and the type of work generally occurring in lane closures and on the shoulder. The proportion occurring in the activity area, advance warning area, and transition area are similar between the two data sets. Despite the higher presence of road work, the road itself does not contribute differently. Table 2 depicts the environmental conditions for Road Ranger crashes as compared with statewide freeway crashes.

Persons

Injuries are more likely when RRSP are involved. Of RRSP crashes, 37% result in at least one injury, whereas only 27% of Florida freeway crashes cause injury. Because of the small sample size, fatalities are difficult to compare, but there were two fatal RRSP crashes during the period, representing 1%. Fatal freeway crashes account for just 0.4% of the more than 176,000 crashes in that set.

Drugs, alcohol, and distraction all increase with RRSP crashes. Driver distraction is slightly more likely to be present in the RRSP crash. Alcohol is nearly three times more prominent, and the use of drugs nearly five times more common in RRSP crashes. The FHWA National Responder Training Program notes the dangers of "D" drivers (drunk, drugged, drowsy, distracted), and that is confirmed in the Florida crash data (*16*). Table 3 shows selective data elements and attributes for persons involved in crashes.

Harmful Events

Motor vehicle in transport is typically the first harmful event 81.1% of the time for RRSP crashes and parked motor vehicles 8%. Again, parked motor vehicles are uncommon among general freeway crashes, constituting less than 0.5%. The first harmful event occurs on the roadway at about the same rate, but, as expected, shoulder incidents are more common for RRSP (11% versus 6.7%). Table 4 shows selected variables and attributes for harmful events.

Analysis of Data from Surveys of Road Rangers

Road Ranger operators have the potential to shed light on safety, from the perspective of the individual doing the job. Interviews with FDOT program managers and National SSP contacts provided the basis for survey items. Potential survey questions were circulated to the FDOT program managers for review and comment, and

Table 3. Persons Characteristics Compared

Data element	Attribute	Florida freeway crashes 2014–2016 n = 176,993	RRSP crashes 2014–2016 n = 200
Fatalities	0	99.50%	98.30%
	1	0.40%	1.70%
Injuries	0	73.31%	62.19%
	1	17.43%	26.37%
	2	5.59%	7.96%
	3 +	3.67%	3.49%
Alcohol_related	N	98.28%	95.52%
	Y	1.72%	4.48%
Distraction_related	N	87.41%	83.58%
	Y	12.59%	16.42%
Drug_related	N	99.43%	97.51%
	Y	0.57%	2.49%

Table 4. Harmful Event Comparison

Data element	Attribute	Florida freeway crashes 2014–2016 n = 176,993	RRSP crashes 2014–2016 n = 200
First_harmful_event	Motor vehicle in transport	71.98%	81.09%
	Parked motor vehicle	0.48%	7.96%
	Pedestrian	0.18%	3.48%
First_HE_location	Median	3.00%	1.00%
	Off roadway	6.11%	6.47%
	On roadway	83.73%	79.60%
	Shoulder	6.74%	10.95%
First_HE_relation_to_jct	Acceleration/deceleration lane	0.12%	0.50%
	Driveway/alley access related	0.02%	0.50%
	Entrance/exit ramp	11.25%	9.95%
	Non-junction	82.92%	78.11%
First_HE_within_interchange	N	91.54%	88.56%
	Y	8.19%	10.95%

the final survey instrument was reviewed and approved by the FDOT statewide Traffic Incident Management (TIM) Program Manager. The survey contained a total of 33 items, including demographic information, vehicle and equipment, training, operations and procedures, and driver perceptions topics.

The survey instrument was distributed as both a PDF file and a web link to FDOT RRSP program managers, who in turn solicited their RRSP contractors to have drivers complete the survey. Surveys were collected by the FDOT RRSP Program Manager and then returned to the research team. The participation in the survey was exceptional with 217 drivers participating, which is very high given that there are 105 unique RRSP beats (*17*).

Demographic Information

Employment information about the driver taking the survey captures the FDOT District where they work, which type of RRSP vehicle they drive, how many years they have been working, and whether the work mostly during daytime or nighttime hours.

All FDOT districts currently use Road Rangers and all participated in the survey. According to survey responses, many RRSP drivers have been on the job for more than 5 years (45%). On the opposite end of the continuum, 26% have been on the job for 1 year or less. This interesting dichotomy of experience has potential to be enlightening, and will be explored with a series of cross tabulations in the analysis section later in this paper.

SSP were created to deal with peak hour traffic on urban freeways, typically daytime hours. Over time, the value of SSP has spread to off-peak hours and even 24-h coverage. Days and hours of operation are dictated at the District level, particularly in contracting agreements. Asking drivers if they work mostly during daytime or nighttime hours has value because darkness presents unique safety challenges.

Vehicle and Equipment

Vehicles and equipment were evaluated using several survey items. Minimum equipment required for RRSP contractors is set forth by contract, so querying details about those items is not of value. The objective of the questions about vehicles and equipment is to consider vehicle markings, lighting, arrow boards, advance warning and high-visibility safety apparel (HVSA) sufficiency.

Several FDOT RRSP program managers indicated that tow-capable vehicles are a valuable resource. Unfortunately, budget constraints have created circumstances in which towing vehicles have been replaced with less expensive pickup trucks. When operated as an RRSP, tow trucks are dedicated and marked as full-time Road Rangers. Whereas several drivers indicated that they use both pickup truck and tow truck, more than two thirds of RRSP operators typically drive a pickup truck.

The sufficiency of RRSP vehicle conspicuity markings and emergency lighting were captured using a graphic rating scale. Survey respondents were asked to rate these items as "excellent," "good," or "needs improvement." Most drivers viewed conspicuity markings on the rear of the RRSP vehicle as "excellent," and nearly nine in 10 felt them to be "excellent or good."

A similar question about emergency lighting was answered quite differently, with 27.9% of drivers indicating that that was an area of equipment that required improvement. Many drivers actually commented on vehicle emergency lighting in responses to open-ended questions about equipment and safety. Survey respondents felt strongly about the relationship between emergency lighting and safety with 70% indicating they felt that more lights would enhance safety.

The Manual on Uniform Traffic Control Devices (MUTCD) designates that a fluorescent pink advance warning sign is reserved for TIM applications (*18*). Though not required by FDOT contracts, many RRSP carry these signs on their vehicle to augment arrow boards and other temporary traffic control devices. According to the survey, 26% of operators carry the fluorescent pink advance warning sign.

Florida RRSP use and monitor the Florida State Law Enforcement Radio System (SLERS). The use of SLERS is beneficial given that the Florida Highway Patrol (FHP) is responsible for incidents on Florida freeways and also because FHP dispatch is co-located with the FDOT TMC. Drivers can communicate with the TMC and other responders about incidents. According to the survey, 88% of drivers have a SLERS radio, 53% monitor the radio, and 88% believe it is important to their safety. Drivers were asked if communications coverage (radio or cellular) was ever a problem and just over 70% responded affirmatively, making the use of SLERS even more important.

HVSA, commonly known as high-visibility traffic vests, are required by the MUTCD, and by FDOT RRSP contracts. Every RRSP driver is required to use a MUTCD-compliant, American National Standards Institute (ANSI) traffic vest at all times. Drivers were questioned about the potential value of other HVSA, such as pants, to improve responder safety. A total of 64% of drivers felt that adding high-visibility pants would improve operator visibility. As a related note, many drivers mentioned traffic vests and uniforms in open-ended questions, indicating a desire for uniform options which combine both.

Training

Road Ranger contractors bear the responsibility for training RRSP driver/operators. FDOT procedures specify basic first aid, CPR, Intermediate MOT (Maintenance of Traffic), and SLERS radio as the minimum training for operators. Clearly there are many competencies required of the job, so additional training is accomplished with on-the-job training. For Florida, the number of contractor training hours varies, so questions focusing on topical competencies are a better gauge of sufficiency. Using a graphic rating scale, drivers were asked to rate their training in several common areas as "excellent, good, or needs improvement." Table 5 depicts responses from drivers.

Table 5. Road Ranger Operator Responses to Training Question

Training topic	Excellent		Good		Needs improvement		Total
Push bumper procedures	61.40%	132	30.70%	66	7.91%	17	215
Debris procedures	55.81%	120	36.28%	78	7.91%	17	215
Spill procedures	50.47%	107	41.98%	89	7.55%	16	212
Fire extinguisher	61.86%	133	33.49%	72	4.65%	10	215
Crash procedures	68.37%	147	30.23%	65	1.40%	3	215
Disabled vehicle procedures	68.22%	146	30.37%	65	1.40%	3	214
Temporary traffic control of MOT	68.69%	147	28.04%	60	3.27%	7	214

More than half of drivers rated their training in all topical areas as "excellent," a good indication that the current contractor delivery is useful to drivers. Push bumper procedures, debris procedures, and handling spills were areas in which drivers indicated there was a need for improvement, but by fewer than one in 10 survey participants. Another encouraging sign about training was that, when asked if they had "all of the training and equipment needed to do their job," 95% of respondents answered affirmatively.

Recurring training is important to drivers remaining current. Because many drivers have years of experience, continuous training is important. To identify how widespread monthly training was, drivers were asked if they participated in some type of monthly training. Approximately half of drivers have monthly training.

The national standard for incident management training is the FHWA National TIM Responder Training Program. Developed under the second Strategic Highway Research Program (SHRP2), through July 2017, more than 260,000 police, fire, emergency medical service, transportation, and towing professionals have attended the 4-h in-person class or taken the equivalent online program. A total of 59% of Florida RRSP operators have taken the training, according to survey results.

Operations and Procedures

Vehicle operation is an area in which risk for RRSP operators is increased. Under authority granted in Florida Statute 316.003(1), the FDOT has designated RRSP vehicles as authorized emergency vehicles. This allows drivers of RRSP vehicles to disregard regulations governing parking, direction, turning, movement, and so on under certain circumstances (*19*). Despite the legal authority, some RRSP contractors may impose company policies that restrict the use of shoulders and median openings for insurance reasons. According to the survey, 22% of drivers use median turnarounds on routine patrol, 45% use median turnarounds when responding to an incident, and 81% use the roadway shoulder for response to incidents.

A series of questions sought to understand exposure that RRSP operators have in hazardous situations. RRSP operators indicated that they "often" (47%) or "sometimes" (41%) work on bridges or elevated roadways where there is limited space. A follow-up question sought to see if they had assistance from other responders in those situations, providing advance warning or a blocking vehicle. When operating in those conditions, 37% of drivers "often" had help, 34% "sometimes," and 29% "rarely."

A basic principle of responder safety is avoiding standing near active lanes of traffic. Drivers were asked if they use a technique called an "off side approach" when engaging disabled motorists or others. In this technique, the RRSP operator would approach and stand for example on the right side of a car on the right shoulder, speaking with the occupant(s) through the passenger window. Nine out of 10 RRSP drivers indicated that they "often" use this technique when making initial contact, 6% "sometimes," and 4% "rarely." A follow-up question sought to understand exposure in situations in which the RRSP operator may be confronted with working near moving traffic. When asked how often they provide fuel or change tires on the traffic side of a disabled vehicle, 66% said "often," 22% "sometimes," and 12% "rarely."

Perceptions and Opinions

Driver opinions and perceptions are a valuable part of safety analysis, because they accommodate topics that may have not been adequately covered in structured questions. In open-ended questions, drivers were noticeably concerned about "move over" behavior among drivers who pass their incident scenes. The drivers commented that more enforcement was needed to improve compliance, but also wrote that additional emergency lights or different color lights would improve move over.

A related question centered on the use of vehicle-mounted arrow boards while on the shoulder. The MUTCD specifies that while on the shoulder, arrow boards should be placed in the "caution" mode, in which lights on the board simply display a dot in each of the four corners of the display (*18*). Incident responders have often desired to use an arrow on the display board to promote move over behavior. Drivers were asked if they felt using the arrow would improve move over compliance and 72% said "yes," with an additional 11% "not sure," and the remaining 17% indicating "no."

Vehicle-mounted digital message boards are an alternative to the vehicle-mounted arrow board on most RRSP. Digital message boards, sometimes called changeable message boards, allow the operator to display text messages or graphics like an arrow. RRSP drivers were asked if they thought message boards were superior to simple arrow boards and 47% indicated "yes," 31% "no," and the remaining 22% "not sure."

Work zones are difficult areas of work for SSP, particularly where there are reduced or non-existent shoulders in work zones. Drivers were asked to rate the importance of "shoulders or crash investigation sites in work zones." All drivers thought they were "important" or "very important."

The analysis presented thus far focused on the entire set of Road Rangers who responded to the survey. Given that the responses may depend on various characteristics

Table 6. RRSP Operator Experience Cross Tabulations

Safety question	Response	Experience		Total
		0–5 years	5 + years	
Would additional emergency lights on your RRSP vehicle improve move over compliance?	Missing	0%	1.10%	0.50%
	No	9%	16.80%	12.40%
	Not sure	14.80%	20%	17.10%
	Yes	76.20%	62.10%	70%
	Total	100%	100%	100%
Do you think using the arrow "<=" on the vehicle-mounted arrow board while on the shoulder improves move over?	Missing	0.80%	2.10%	1.40%
	No	13.10%	22.10%	17.10%
	Not sure	9%	12.60%	10.60%
	Yes	77%	63.20%	71%
	Total	100%	100%	100%
Do you think that digital message boards that allow words are better than simple arrow boards?	Missing	0%	1.10%	0.50%
	No	26.20%	35.80%	30.40%
	Not sure	23%	21.10%	22.10%
	Yes	50.80%	42.10%	47%
	Total	100%	100%	100%
When making initial contact with a disabled motorist, how often do you approach their vehicle on the non-traffic side (for example the right side)?	Missing	3.30%	0%	1.80%
	Often	91.80%	83.20%	88%
	Rarely	1.60%	7.40%	4.1
	Sometimes	3.30%	9.50%	6%
	Total	100%	100%	100%

of the Road Rangers, several contingency tables, also called cross tabulations, are developed. In particular the differences in responses are examined based on (1) the experience of the Road Ranger and (2) the shift they work.

When exploring the impact that lighting might have on move over compliance, experienced operators tended to be less optimistic about the potential for improvement. When compared with less tenured operators, fewer experienced operators felt that additional emergency lighting, or the use of arrow boards would improve move over behavior among approaching vehicles. There was also less certainty that the message board is superior to the simple arrow board. On the negative side of behavior, experienced officers tend to use the non-traffic side approach less often. Table 6 lists the cross tabulations between operator experience and several safety questions.

Working in nighttime conditions presents special hazards for all incident responders, particularly RRSP drivers. Table 7 shows that nighttime RRSP operators are about three times more likely to cite a need for improvement in relations with other responders, though the majority still describe relations as "excellent." Additionally, the night shift makes for difficulty in attending safety meetings or training. Drivers who work at night are more inclined to think that enhancing HVSA with pants is a good idea. Drivers who work at night appear to be "not sure" if digital message boards are superior to arrow boards, as depicted in Table 7. Finally, nighttime drivers are highly sensitive to being in travel lanes, ostensibly because of reduced visibility. Whereas all RRSP operators think that shoulders in work zones are important, this is more pronounced for those drivers who work during times of darkness.

Conclusions and Recommendations

Research has indicated rear end collisions are common for RRSP vehicles, and they are struck while parked and on the shoulder at a higher rate than the larger population of freeway crashes. Vehicle conspicuity is very important to mitigate these crash hazards. Passive treatments, such as rear-facing markings on vehicles that include contrasting, fluorescent, and retro-reflective materials, complement emergency lighting and should be used on all RRSP vehicles. Although RRSP operators felt that vehicle markings were sufficient, the topic might be further researched to validate their views.

Compliance with Florida's move over law has the potential to improve safety. Emergency vehicle lighting and the use of vehicle-mounted arrow boards have the potential to improve move over compliance, but these treatments need to be studied further, particularly in the case of the use of arrow boards because they must conform to the MUTCD. Vehicle-mounted arrow boards must be in the "caution mode" while operated on the shoulder, and the use of an arrow or "move over" text may be more effective. This is a change that many operators feel strongly about, but one that would require study to validate those sentiments.

Table 7. RRSP Operator Shift Cross Tabulations

Safety question	Response	Shift assigned		Total
		Daytime	Nighttime	
Do you have monthly safety meetings or training?	Missing	3.9%	4.7%	4.1%
	No	43.8%	56.3%	47.5%
	Yes	52.3%	39.1%	48.4%
	Total	100.0%	100.0%	100.0%
Do you think adding high-visibility pants would improve your personal visibility/safety?	Missing	3.3%	4.7%	3.7%
	No	37.3%	29.7%	35.0%
	Yes	59.5%	65.6%	61.3%
	Total	100.0%	100.0%	100.0%
Do you think that digital message boards that allow words are better than simple arrow boards?	Missing	0.7%	0%	0.5%
	No	32.7%	25.0%	30.4%
	Not sure	15.7%	37.5%	22.1%
	Yes	51.0%	37.5%	47.0%
	Total	100.0%	100.0%	100.0%
How would you rate the importance of including shoulders or crash investigation sites in work zone areas?	Missing	1.3%	0%	0.9%
	Important	26.1%	14.1%	22.6%
	Very important	72.5%	85.9%	76.5%
	Total	100.0%	100.0%	100.0%
Overall, how would you rate the relationships between Road Rangers and other responders?	Missing	1.3%	0%	0.9%
	Excellent	54.9%	51.6%	53.9%
	Good	41.2%	40.6%	41.0%
	Needs improvement	2.6%	7.8%	4.1%
	Total	100.0%	100.0%	100.0%

RRSP operators are more likely to be struck as pedestrians, making situational awareness and the use of HVSA more important. Drivers have indicated that the use of "traffic vests" might be replaced with newer uniform technologies that incorporate MUTCD-compliant ANSI standards in shirts and pants. A review of RRSP uniforms is needed to standardize the approach and take advantage of these technologies. Again, this is an area in which there is need for research. Better HVSA might improve visibility, but also minimize instances in which it is not utilized by accident or for comfort reasons.

RRSP operator training appears to be adequate, from the driver point of view. A formal training needs assessment would identify competencies and guide contractor training, potentially improving safety. The National TIM Responder Training Program is an excellent cross-cutting product for all incident responders, but a specialized training program for SSP may be needed.

Acknowledgments

The research and results which are published in this paper are from a project funded by the Florida Department of Transportation.

Author Contributions

The authors confirm contribution to the paper as follows: study conception and design: Carrick, Srinivasan; data collection: Carrick, Jermprapai; analysis and interpretation of results: Carrick, Srinivasan, Jermprapai; draft manuscript preparation: Carrick, Srinivasan, Jermprapai. All authors reviewed the results and approved the final version of the manuscript.

References

1. Florida Department of Transportation. *Road Ranger Service Patrol*, 2014. http://www.dot.state.fl.us/trafficoperations/traf_incident/rrangers/rranger.shtm. Accessed September 10, 2015.
2. Lin, P. S., A. Fabregas, and H. Zhou. *Review and Update of Road Ranger Cost Benefit Analysis*. BDK, 84, 977-15. Florida Department of Transportation Research Final Report, 2012.
3. Florida Department of Transportation. *Statewide Road Ranger Survey for Incident Responders 2015–2016*. http://www.swfltim.org/Docs/Survey/2016%20Statewide%20Road%20Ranger%20Survey.pdf. Accessed July 26, 2017.
4. National Law Enforcement Officers Memorial Fund. *Cause of Law Enforcement Deaths over the Past Decade (2006–2015)*. The National Law Enforcement Officers Memorial, Washington, D.C., http://www.nleomf.org/facts/officer-fatalities-data/causes.html. Accessed November 27, 2016.
5. Fahy, R. *U.S. Firefighters Killed when Struck by Vehicles, 2000–2013*. National Fire Protection Association. http://www.nfpa.org/~/media/files/news-and-research/fire-statistics/fire-service/osffstruckbyvenicles.pdf. Accessed November 27, 2016.

6. Rodgers, J., and B. Joel. *Towing Cars 'Dangerous Even on a Good Day'*. The Gazette. http://gazette.com/towing-cars-dangerous-even-on-a-good-day/article/113414. Accessed November 27, 2016.
7. Federal Highway Administration. *Facts and Statistics: Work Zone Safety*, 2014. http://www.ops.fhwa.dot.gov/wz/resources/facts_stats/injuries_fatalities.htm. Accessed November 27, 2016.
8. Houston, N., C. Baldwin, A. V. Easton, S. J. Cyra, M. W. Hustad, and K. Belmore. *Service Patrol Handbook*. No. FHWA-HOP-08-031. Federal Highway Administration, U.S. Department of Transportation, Washington, D.C., 2008.
9. Carrick, G., S. Washburn, and S. Srinivasan. Law Enforcement Vehicle Crashes in Florida: Descriptive Analysis and Characterization. *Transportation Research Record: Journal of the Transportation Research Board*, 2010. 2182: 40–47.
10. United States Fire Administration, United States, and International Fire Service Training Association. *Emergency Vehicle Visibility and Conspicuity Study*. Department of Homeland Security, U.S. Fire Administration, Emmitsburg, MD, 2009.
11. Grant, C. C., and B. Merrifield. *Analysis of Ambulance Crash Data*. The Fire Protection Research Foundation, Quincy, MA, 2011.
12. Model Minimum Uniform Crash Criteria. *Guideline: Model Minimum Uniform Crash Criteria*, 4th ed. DOT HS 811 631. US Department of Transportation, Washington, D.C., 2012.
13. Maistros, A., and W. H. Schneider IV. *A Multilevel Statistical Analysis of Impaired Driving Crashes and Law Enforcement Countermeasures*. The University of Akron, Akron, OH, 2015.
14. Savolainen, P., K. Dey, I. Ghosh, T. Karra, and A. Lamb. *Investigation of Emergency Vehicle Crashes in the State of Michigan*. Project 015WY01.1 Final Report. NEXTRANS Center, Region V Regional University Transportation Center, U.S. Department of Transportation, Washington, D.C., 2009.
15. University of Florida. *Signal 4 Analytics*. https://s4.geoplan.ufl.edu. Accessed November 27, 2016.
16. Federal Highway Administration. *Training for Safer, Faster, Stronger, More Integrated Incident Response: National Traffic Incident Management Responder Training Program (L12/L32A/L32B)*. https://www.fhwa.dot.gov/goshrp2/Solutions/Reliability/L12_L32A_L32B/National_Traffic_Incident_Management_Responder_Training_Program. Accessed November 27, 2016.
17. Srinivasan, S., G. Carrick, Y. Yin, K. Jermprapai, X. Sun, N. Shirmohammadi, and M. Shahabi. *Warrants, Design, and Safety of Road Ranger Service Patrols*. FDOT Contract BDV31 977-52. Florida Department of Transportation, Tallahassee, FL, 2016.
18. Federal Highway Administration. *MUTCD 2009: Manual on Uniform Traffic Control Devices*. U.S. Department of Transportation, Federal Highway Administration, Washington, D.C., 2009.
19. Florida State Statute 316.072(5). *Obedience to and Effect of Traffic Laws—Authorized Emergency Vehicles*.

The Standing Committee on Standing Committee on Traffic Law Enforcement (ANB40) peer-reviewed this paper (18-00078).

This paper and its contents, including conclusions and results, are solely those of the authors; they do not represent the opinions or policies of the funding agency.

Article

TRR
JOURNAL OF THE TRANSPORTATION RESEARCH BOARD

Transportation Research Record
2018, Vol. 2672(33) 42–55

DOI: 10.1177/0361198118778943
journals.sagepub.com/home/trr
SAGE

Evaluating Wrong-Way Driving Characteristics, Countermeasures, and Alert Dissemination Methods through Driver and Law Enforcement Surveys

Sara Wertanen[1], Corin Staves[1], Haitham Al-Deek[1], Adrian Sandt[1], Grady Carrick[2], and John H. Rogers, Jr[1]

Abstract

Wrong-way driving (WWD) on limited-access highways is a serious driver error that could be fatal for both the wrong-way driver and right-way drivers. Agencies and law enforcement officers (LEOs) can spend valuable time and resources responding to WWD events. A survey of 247 LEOs found that despite being dispatched multiple times per year in response to WWD, most issue fewer than one WWD citation per year. In the survey, LEOs also gave their perspectives on the characteristics of WWD events and the influences behind WWD. Agencies typically use dynamic message signs (DMS) for notifying right-way drivers of WWD events, but little research has considered solutions that are less costly and potentially farther-reaching. A survey of 900 Florida toll-road drivers found that drivers rely on navigation devices, apps, or both, significantly more than DMS for traveler information. Drivers were receptive to receiving WWD alerts through navigation services and indicated that they would react to WWD events in a way that is beneficial to highway safety. Crowdsourcing technologies were considered to generate WWD information and quickly disseminate alerts to drivers and officers. Additional survey questions answered by a combined group of 2,052 drivers, LEOs, and agency officials determined that among advanced WWD countermeasures, most preferred two sets of "Wrong Way" signs equipped with rectangular flashing beacons over one set of signs with light emitting diodes. Combining advanced countermeasures with innovative right-way driver notification methods could enhance law enforcement efficiency, reduce WWD crashes, and provide safer conditions for both drivers and officers.

Although wrong-way driving (WWD) is relatively rare on limited-access highways, wrong-way crashes are often fatal and receive significant media attention. From a law enforcement perspective, responding to WWD events takes up valuable time and resources. As part of ongoing work in the field of WWD, University of Central Florida (UCF) conducted a law enforcement survey to better understand how their resources are allocated when addressing WWD. The 247 Law Enforcement Officers (LEOs) who completed the survey were asked to compare the frequency at which they encounter WWD crashes, 911 call dispatches, and citations, and to provide their perspectives on the characteristics and causes of WWD. Understanding these factors can help LEOs and traffic agencies effectively detect, alert, and stop wrong-way drivers.

Although WWD countermeasures are effective at alerting and stopping wrong-way drivers, some of those drivers will continue the wrong way, often with tragic consequences. Therefore, agencies need to determine the best methods to notify right-way drivers of detected WWD events. Agencies primarily use dynamic message signs (DMS) for these notifications. However, DMS can be sparsely located, even on urban freeways. Alternatives are needed to fill the gaps in DMS coverage in areas where additional DMS signs would be cost prohibitive or not appropriate. Many in-vehicle navigation assistants are Internet-connected and provide real-time traffic information directly to drivers. Furthermore, smartphones have become ubiquitous in the modern world, with many drivers relying on apps for navigation and traffic updates. Utilizing these technologies could allow agencies to provide WWD alerts to a large portion of travelers in a short time. Connected vehicles could also be programmed to detect WWD and alert drivers. However, though connected vehicle technology is still in its infancy, it is important to realize that many solutions exist today to keep drivers connected and informed. If WWD notifications are incorporated into existing communication and navigation technologies, they could potentially be received by a much larger percentage of drivers

[1]Department of Civil, Environmental, and Construction Engineering, University of Central Florida (UCF), Orlando, FL
[2]Enforcement Engineering, Inc., Saint Johns, FL

Corresponding Author:
Address correspondence to Haitham Al-Deek: Haitham.Al-Deek@ucf.edu

in danger. The delay between a wrong-way driver being reported and right-way drivers being notified would also decrease, further reducing the likelihood of a crash. To understand which WWD alert dissemination methods would be the most effective, a survey was conducted with 900 Central Florida toll-road drivers.

The information obtained from this driver survey could also help LEOs. Pursuing wrong-way drivers can be a dangerous task for officers, especially when they must travel the wrong way themselves in pursuit of a wrong-way vehicle (*1*). Therefore, it is imperative that LEOs know as much as possible about the wrong-way vehicle so they can locate and stop it before a crash occurs. A previous UCF study found that only 10% of drivers who had witnessed WWD on a limited-access highway reported it by calling 911 or notifying the appropriate agency (*2*). Crowdsourced navigation and reporting services provide an alternative method for agencies and drivers to obtain fast and precise WWD information. The smartphone application Waze enables users to report traffic jams, crashes, and other roadway incidents in real-time to alert other drivers. If WWD events could be reported through services like Waze, then both LEOs and other drivers could be alerted to WWD more frequently, more quickly, and more accurately, even in areas without advanced detection infrastructure. Survey questions given to 319 Waze users were used to evaluate the viability of WWD reporting in Waze while considering possible safety implications.

By obtaining LEO perspectives on the characteristics and causes of WWD and driver perspectives on notification methods, this study could help agencies determine which WWD countermeasures might be the most effective. An increasingly popular countermeasure is to add flashing lights to "Wrong Way" signs on highway exit ramps. When a wrong-way driver is detected, the lights begin flashing to alert the driver so that they turn around before entering the mainline. Variations of this countermeasure have been implemented throughout Florida. The Florida's Turnpike Enterprise (FTE) in South Florida implemented "Wrong Way" signs with light emitting diodes (LEDs) around the edge on either side of the exit ramp. These LEDs blink in a constant pattern when a wrong-way vehicle is detected. In Central Florida, however, the Central Florida Expressway Authority (CFX) and FTE have implemented a different solution that adds a second set of signs further up the ramp and utilizes rectangular flashing beacons (RFBs) instead of LEDs. With RFBs, two large lights are placed below the sign and another two lights above the sign that flash in an alternating pattern. These devices were previously called rectangular rapid flashing beacons (RRFBs), but are now called RFBs to differentiate them from RRFBs used at pedestrian crossings, which have a different flashing pattern and frequency.

This study analyzed 2,052 responses from four different surveys to determine which method (the one set of LEDs or the double set of RFBs) is preferred by drivers, LEOs, and Florida Department of Transportation (FDOT) agency officials.

Before detailing the survey results, a review of previous research on WWD detection, countermeasure, and notification systems, as well as examples of systems that are currently used throughout the country is discussed. This literature review also introduces alternative notification systems and discusses their adaptability to WWD. Then, the results from the law enforcement survey are presented and discussed in detail, followed by the results from the Central Florida driver survey. Lastly, the questions regarding the two advanced WWD countermeasure systems currently implemented in Florida are discussed using responses from four surveys. The results from all these surveys will assist agencies in providing the most effective WWD notification and countermeasure solutions to keep drivers and officers safe.

Previous Research on WWD Countermeasures, Alert Dissemination Methods, Warning Systems, and Driver Surveys

Although WWD collisions account for only about 3% of crashes on freeways, state studies have shown their fatality rate to be between 12 and 27 times higher than other highway crashes (*3*). Owing to the increased media exposure of WWD crashes, research into WWD detection and countermeasure systems has become increasingly popular. For example, many states have improved the ubiquitous "Wrong Way" and "Do Not Enter" signs on freeway exit ramps by providing larger signs, lower placement, reflective tape, additional signs or additional markings or both (*4*, *5*). States with especially high rates of WWD events, such as Florida and Texas, have implemented advanced intelligent transportation system (ITS) detection and countermeasure solutions, improving "Wrong Way" signs with flashing LEDs that are set off by radar or laser detectors on the side of the road (*1*, *6*, *7*). For example, FDOT District-7 implemented a single set of RFBs on 20 interstate ramps in and around Tampa, Florida (*8*). Texas Department of Transportation (TxDOT) implemented an earlier generation of LED sign technology on their San Antonio corridor (*9*). However, the double set of RFBs used in Central Florida has received national attention as a potentially more effective countermeasure (*7*, *10*). The success of any WWD countermeasure depends largely on agencies' knowledge of the characteristics of wrong-way drivers. In an National Transportation Safety Board analysis of 1,150 WWD crashes, 69% of the wrong-way drivers had a blood alcohol content above the legal limit (*3*).

Despite the growing popularity of advanced WWD detection and countermeasure systems, little research has taken place with respect to notifying right-way drivers of detected nearby wrong-way vehicles. Previous simulation work showed that such notifications are beneficial to highway safety as drivers who receive them are likely to move over to the right lane (*11*). One of the earliest right-way driver notification solutions, implemented before 2004, involved flashing lights on a span wire activated by directional loop detectors on a particularly WWD-prone bridge in Pensacola Bay, Florida (*12*). In Rhode Island, Texas, and Florida, DMS are currently used to warn right-way drivers about a nearby wrong-way vehicle (*7*, *9*, *13*). DMS signs have proven to be a popular solution among drivers; a UCF study in 2009 determined that 90% of respondents found DMS to be helpful in warning about hazardous situations (*14*). However, it has been pointed out that a shortcoming of DMS is their spacing and limited reach, especially in rural areas, which leads to significant gaps in coverage (*15*). Therefore, blank-out signs were considered (but never implemented) for notifying right-way drivers in rural areas where additional DMS infrastructure would be impractical (*16*). Blank-out signs are small electronic signs that display a single static message when activated and are otherwise blank.

Wireless emergency alerts (WEA) have also been considered as a possible tool to "broadcast the location and direction of a wrong-way vehicle" (*16*). WEA is a public safety system that sends geographically targeted emergency warnings to eligible devices (*17*). With current technology, the warning zone is defined by cell tower locations and alerts are broadcast to every device within range of the specified tower(s). This causes alerts to "bleed over" into surrounding areas and makes it difficult to define alert zones with a high level of precision (*18*). However, the Federal Communications Commission is currently investigating new WEA technology, called arbitrary-size location-aware targeting (ASLAT), that will allow alerts to be broadcast with much greater precision (*19*). With ASLAT, alerts include non-displayable coordinates of the warning target zone, and the targeted mobile phones use their own geolocation technology (such as GPS) to decide whether to display the alert. In addition to geographic coordinates, this non-displayable information can also include details on trajectory, which would be useful for restricting alerts to drivers traveling in a specific direction (*19*).

Connected vehicles are emerging in the automotive industry as a means of creating a safer driving experience. These connected vehicles can communicate various aspects of traffic and roadway information to other surrounding vehicles. Therefore, connected vehicle technology has potential to be used for WWD notifications. A previous UCF survey found that 91% of participants believed drivers should receive WWD warning messages through in-vehicle warning systems, which could be achieved using connected vehicle technology (*7*). Texas A&M Transportation Institute aimed to create a "concept of operations, functional requirements, and high-level system design for a Connected Vehicle WWD Detection and Management System" (*20*).

Since connected vehicle technology is not currently commonplace, other technologies need to be considered. Waze is a popular smartphone navigation app with crowdsourced traffic and incident reporting features that transmit alerts in real-time between users. Users can report data such as road hazards and traffic warnings to notify other drivers. A points system is used to give drivers an incentive to report events. An online statistic reported that Waze received an event before a 911 call was placed 62% of the time (*21*). Agencies have realized the potential of Waze, with FDOT being among the first transportation agencies nationally to partner with Waze, sharing data between their traffic management centers (TMCs) and the application provider. Real-time Waze information for Florida freeways is absorbed into the FDOT TMC software and presented as alerts to operators who are co-located with Florida Highway Patrol (FHP) dispatchers (*22*). LEOs have also shown support for Waze as it can urge users to drive more carefully (*23*).

A review of literature shows that previous studies on notifying right-way drivers of WWD have focused mainly on DMS. However, technologies such as WEA and navigation services could be adapted to provide WWD alerts. Crowdsourced technologies such as Waze could be used to share and disseminate information about the presence of a wrong-way vehicle. By examining the viability of these solutions, this research will provide useful knowledge for agencies and highway authorities that has not been previously obtained. This paper will also summarize the perspectives of road users, law enforcement, and public traffic agencies regarding WWD countermeasures. With this information, agencies will be able to better gauge which WWD countermeasures and alert dissemination methods could maximize user safety and experience.

Survey Methodology

The main components of this research include four online-conducted surveys. The completed surveys were analyzed to determine the most common responses (mode) for each question. Frequency analyses were performed to find important relationships between driver demographics, reactions to WWD notifications, and technology preferences.

FHP Law Enforcement Survey on WWD Characteristics and WWD Countermeasures

This survey was conducted on 247 FHP officers. FHP is an important agency for WWD response, since they have primary patrol responsibilities for Florida freeways and toll roads and they currently investigate more than 90% of crashes occurring on these limited-access highways. The survey was designed to understand the extent of WWD from a law enforcement perspective and to obtain information on potential WWD causes and countermeasures. The survey was emailed to FHP personnel and completed by officers using an online form. As a screening criteria, only officers who had previously worked on at least one WWD event (defined as a crash, citation, or dispatch) could complete the survey.

CFX Driver Survey on Notifying Right-Way Drivers and WWD Countermeasures

This survey was conducted on 900 random toll-road users in Central Florida. This survey primarily focused on WWD alert dissemination methods, with an emphasis on the smartphone application Waze. In this survey, drivers were asked about their current information dissemination technology choice, their opinions on various WWD alert methods, their reactions to WWD alerts, theoretical WWD Waze-reporting questions, WWD countermeasure questions, and social demographic questions. Respondents were randomly chosen from a panel of professional survey respondents. All respondents were toll-road users of the CFX network.

FTE Driver Survey Questions on WWD Countermeasures

This survey was conducted on 900 random FTE toll-road customers and interstate users in South Florida. Details of this survey were previously discussed in Sandt et al. (*7*). Only the results from the WWD countermeasure questions were included in this paper; these were combined with the other surveys' results.

FDOT Agency Survey Questions on WWD Countermeasures

This survey was conducted on FDOT stakeholders. Like the FHP survey, this was emailed to FDOT representatives from the eight FDOT districts and completed using an online form. Five districts responded. As it is hard to make definitive conclusions from this survey with such a small sample size, only the results from the WWD countermeasure questions were included and combined with results from the other surveys.

Survey Results

FHP Law Enforcement Survey

The questions, response options, and response percentages for the FHP survey are shown in Table 1. Percentages were rounded to the nearest whole percent. Due to rounding, percentages might not add up to exactly 100%. For each question, the most popular response (mode) is shown in bold. There was one FHP troop designation question, three WWD experience questions and 10 WWD characteristics questions. This survey also included four WWD countermeasure questions, which are discussed in a later section. Every geographic area of the state was represented in the survey, with Troop C (central-west Florida, includes Tampa Bay area), Troop L (south-east Florida, includes West Palm Beach and Fort Lauderdale), and Troop D (central-east Florida, includes Orlando and Daytona Beach) well represented.

WWD Experience Questions. The objective of these questions was to better understand the extent of WWD activities for LEOs. It was found that 73% of officers respond to at least one WWD crash per year. Even more (81%) respond to at least one noncrash WWD event per year. However, only 40% of officers issued one or more WWD citations per year. It was discovered that even though most officers respond to multiple WWD events per year, it is difficult for them to locate and stop wrong-way drivers before they disappear or self-correct.

WWD Characteristics Questions. The main focus of the law enforcement survey was the characteristics of WWD events and wrong-way drivers. For most FHP officers (73%), the WWD events that they work on occur mainly in urban or suburban areas. Furthermore, most officers (62%) said that over 60% of the wrong-way dispatch events they respond to are either "never found, never pulled over, or never crashed." This is consistent with the results of the experience questions. About 83% of respondents said it is very difficult, difficult, or somewhat difficult to determine where the wrong-way driver originated. This shows that determining the exact interchange where a wrong-way driver enters a limited-access facility can be challenging. Most officers (90%) have worked on cases where the wrong-way driver traveled at least 1 mi before being stopped, and 47% have worked on cases where the driver traveled at least 4 mi. This indicates that wrong-way drivers can travel far before either self-correcting, being stopped by LEOs, or causing a crash.

To understand the characteristics of wrong-way drivers, officers were asked about various influences associated with WWD citation events. These influences were ranked using a graphical rating scale of 1 (not common) to 4 (very common) and respondents were asked to provide separate rankings for daytime and nighttime citations. Driving under the influence (DUI) was the most common influence in nighttime citations, with an average rating of 3.6. This is consistent with previous research which has shown that DUIs are the most significant cause of WWD. For daytime events, the influence of DUI was significantly lower at 2.7. Confused and elderly drivers were the most common influence in daytime citations, with an average rating of 2.9. The average daytime and nighttime ratings for each influence are shown in Figure 1.

DUI citations are common for wrong-way drivers, as 77% of officers stated that they are commonly issued to wrong-way drivers. Most officers (96%) said that wrong-way drivers typically travel alone, and 40% of officers said that it can be difficult or very difficult to determine where an intoxicated driver started his or her trip. The surveyed officers rarely recommend drivers cited for WWD for re-examination, which is reserved for the most serious cases where the driver might have diminished cognitive or other driving skills.

Table 1. Law Enforcement (FHP) Agency Survey Questions and Responses

Question category (type)	Question	Response choices	Response percentages
Troop designation (categorical)	Your FHP troop designation	A	7%
		B	5%
		C	**16%**
		D	13%
		E	7%
		F	8%
		G	7%
		H	8%
		I	3%
		J	1%
		K	10%
		L	**16%**
Experience (ordinal)	How many wrong-way crashes do you typically respond to in the span of a year?	Less than one a year	27%
		One (1) a year	16%
		Two (2) to three (3) a year	**34%**
		Four (4) to five (5) a year	6%
		More than five (5) a year	16%
Experience (ordinal)	How many wrong-way citations, outside of crash events, do you typically issue in the span of a year?	**Less than one a year**	**60%**
		One (1) a year	16%
		Two (2) to three (3) a year	17%
		Four (4) to five (5) a year	2%
		More than five (5) a year	4%
Experience (ordinal)	How many wrong-way dispatch events, outside of crash events, do you typically respond to in the span of a year?	Less than one a year	19%
		One (1) to three (3) a year	**33%**
		Four (4) to seven (7) a year	27%
		Eight (8) to ten (10) a year	8%
		More than ten (10) a year	13%
Characteristics (categorical)	What other citations do you commonly couple with wrong-way citations?*	Reckless driving	13%
		Careless driving	19%
		Driving under the influence	**77%**
		Other	9%
Characteristics (categorical)	Where do most of the wrong-way events (crashes, citations, and dispatch calls) that you work on occur?	**Urban area interstates/ freeways**	**44%**
		Urban area toll roads	7%
		Suburban area interstates/ freeways	19%
		Suburban area toll roads	3%
		Rural interstates/freeways	24%
		Rural toll roads	3%
Characteristics (categorical)	Based on your experience, are most wrong-way drivers that you have encountered alone in their vehicles?	**Yes**	**96%**
		No	4%
Characteristics (categorical)	Please score the following influences that are associated with nighttime WWD citation events, excluding crashes	(Multiple sub-questions)	See Figure 1
Characteristics (categorical)	Please score the following influences that are associated with daytime WWD citation events, excluding crashes	(Multiple sub-questions)	See Figure 1
Characteristics (ordinal)	With respect to wrong-way dispatch events that you have worked, what percentage of these events are gone on arrival and are either never found, never pulled over and/or never crashed?	0% to 20%	9%
		21% to 40%	11%
		41% to 60%	18%
		61% to 80%	**33%**
		81% to 100%	29%

(continued)

Table 1. (Continued)

Question category (type)	Question	Response choices	Response percentages
Characteristics (ordinal)	When working a wrong-way crash, how difficult is it to determine where the wrong-way driving started or entered the roadway?	Not difficult	17%
		Somewhat difficult	**43%**
		Difficult	29%
		Very difficult	11%
Characteristics (ordinal)	When any wrong-way crash or citation event involves a driver under the influence, how difficult is it to determine where they may have had their last drink or initiated their trip?	Not difficult	15%
		Somewhat difficult	**45%**
		Difficult	32%
		Very difficult	8%
Characteristics (ordinal)	With respect to wrong-way citations that you have issued, outside of crashes, what percentage of these drivers have you recommended for driver re-examination?	**0% to 25%**	**74%**
		26% to 50%	13%
		51% to 75%	5%
		76% to 100%	8%
Characteristics (ordinal)	In wrong-way events that you have worked, what is the longest known distance a driver has traveled in the wrong direction?	Less than 1 mile	10%
		One (1) to three (3) miles	**43%**
		Four (4) to seven (7) miles	26%
		Eight (8) to ten (10) miles	9%
		More than ten (10) miles	13%

Note: Mode in bold.
*Respondents could select multiple responses.

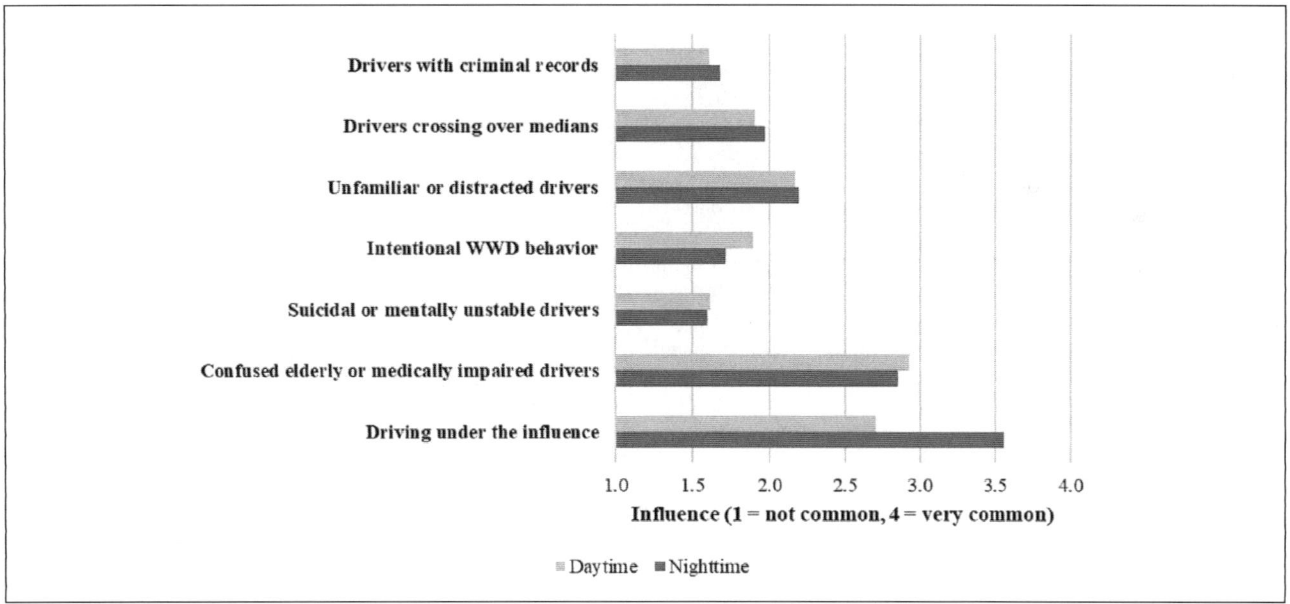

Figure 1. Influences associated with WWD citation events.

CFX Driver Survey

The questions, response options, and response percentages for the CFX driver survey are shown in Table 2. Percentages were rounded to the nearest whole percent. Due to rounding, percentages might not add up to exactly 100%. For each question, the most popular response (mode) is shown in bold. There were five traveler information dissemination technology choice questions, eight opinion questions on alert dissemination methods, two reaction questions, and three Waze-reporting questions. This survey also included three WWD countermeasure questions, which are discussed in a later section. Before the survey

Table 2. CFX Notifying Right-Way Drivers Survey Questions and Responses

Question Category (Type)	Question	Response Choices	Response Percentages
Filtering (categorical)	Please enter your age.	18–25	12%
		26–35	17%
		36–50	**25%**
		51–65	24%
		65 over	23%
Information Dissemination Technology Choice (categorical)	Do you use or have you ever used Waze?	Yes	35%
		No	**65%**
Information Dissemination Technology Choice (ordinal)	How often do you use a navigation assistant, such as a GPS device or a navigation app?	**Frequently**	**52%**
		Sometimes	36%
		Rarely	8%
		Never	5%
Information Dissemination Technology Choice(categorical)	Do you connect your phone to your car via USB, Audio Jack, or Bluetooth?*	USB	29%
		Audio Jack	19%
		Bluetooth/Wireless connection	**51%**
		I do not connect my phone to my car	29%
Information Dissemination Technology Choice (categorical)	Which of the following do you rely on most for receiving travel information while driving?*	DMS Signs	30%
		Highway Advisory Radio (HAR)	9%
		Personal Navigation Device	**61%**
		Traffic Reports on Commercial Radio	32%
		Smartphone Applications	57%
		Web Browser/Search Engine	10%
		511 Phone Service	3%
Opinion (categorical)	Would you want to be warned of a wrong way driver detected on roadways near where you are traveling?	**Yes**	**97%**
		No	3%
Opinion (categorical)	Do you think blank-out signs should be installed on the side of the highway that display a warning message about any detected wrong way vehicle on the next ramp?	**Yes**	**90%**
		No	10%
Opinion (categorical)	Do you think wireless emergency alerts should be used for wrong way driver notification?	**Yes**	**74%**
		No	26%
Opinion (ordinal)	What is the maximum range you think the wrong-way driving emergency alert should be broadcast?**	2 miles	17%
		5 miles	**39%**
		10 miles	26%
		15 miles	8%
		25 miles	7%
		50 miles	2%
		100 miles or more	1%
Opinion (categorical)	Do you think that advanced methods of WWD detection (such as cameras or radar detectors at off-ramps) should automatically create an alert in Waze?***	**Yes**	**94%**
		No	6%
Opinion (categorical)	Do you think navigational assistants should have real time alerts when a wrong way driver is detected?	**Yes**	**93%**
		No	7%
Opinion (categorical)	How would you like to receive a notification of a wrong way vehicle?*	Text Message through Cell Phone	38%
		Wireless Emergency Alert	63%
		Radio Broadcast	37%
		Navigational Assistance Alert	**63%**
		In-Vehicle Alerts	36%
		Overhead Dynamic Message Sign	51%
		Would not want to receive notifications	1%

(continued)

Table 2. (Continued)

Question Category (Type)	Question	Response Choices	Response Percentages
Opinion (ordinal)	How receptive would you be to the concept of vehicles electronically passing warning information between each other along a section of highway to warn motorists of a wrong way driver (this concept is also known as 'Connected Vehicles')?	**Completely receptive**	**41%**
		Somewhat receptive	31%
		Neutral	21%
		Somewhat unreceptive	4%
		Not receptive at all	3%
Waze Reporting (ordinal)	If you saw a wrong way driver while traveling, how likely would it be for you or your passenger to report it in Waze?***	***Would definitely report it***	**71%**
		Probably would report it	23%
		Probably would not report it	5%
		Would not report it	2%
Waze Reporting (categorical)	How would you report the wrong way driver in Waze?***/**	Right away while driving	25%
		Pull over and report it ASAP	**41%**
		Exclusively have the passenger report it	35%
Waze Reporting (categorical)	Would you be more likely to report a wrong way driver in Waze if a higher point value was assigned to this type of alert?***	**Yes**	**61%**
		No	39%
Reaction (categorical)	If you were warned of a wrong way driver approaching you on a toll road or expressway (regardless of how the warning was delivered or any other information), what would you do?	Continue driving in current lane	6%
		Move over to the right lane	21%
		Move over to the middle lane	3%
		Move over to the left lane	3%
		Pull over and stop on right shoulder	**39%**
		Pull over and stop on left shoulder	2%
		Exit the toll road or expressway	26%
Reaction (ordinal)	How long would you wait on the shoulder until you would continue driving?**	Less than 2 minutes	4%
		2 minutes to 5 minutes	24%
		5 minutes to 10 minutes	15%
		More than 10 minutes	1%
		Until you receive more information	**56%**

Note: Mode in bold.
*Respondents could select multiple responses.
**Only asked to respondents who gave the italicized response in the previous question.
***Only asked to Waze users.

began, potential respondents were filtered based on age. Quotas were set for each age bracket, so the age distribution of the survey respondents approximated the age distribution of Florida. After filtering, 900 random respondents completed the survey. Demographic information was also collected regarding gender, educational level, and household income. Except for those with a household income of over US$100,000 and those with an educational level of high school diploma or less (who were both underrepresented in this survey), it was found that the demographic distribution of the survey respondents closely represented the demographic distribution of the entire United States.

Current Traveler Information Dissemination Technology Choice Questions. To determine the best delivery method for disseminating WWD notifications, respondents were asked about how they currently receive traffic information and how they connect their phone to their vehicle. Most survey respondents connect their phone to their car, with 51% utilizing wireless connections (like Bluetooth). Drivers would prefer to receive travel information via navigation assistants, specifically personal navigation devices (PNDs) (61%) and smartphone apps (57%). These are especially popular among drivers aged 18 to 50, as seen in Figure 2*a*. Although it is important to continue utilizing all currently available traffic information systems to distribute traffic warnings, these results show that travelers rely on PNDs and apps more than DMS (30%) or other notification methods.

Figure 2*b* shows that regardless of the income bracket, over 80% of respondents use navigation assistants. This suggests that wrong-way driver alerts distributed through navigation assistants would be equally beneficial to drivers of all incomes. However, people with higher income are more likely to use Waze, as shown in Figure 2*c*. This suggests that Waze usage will be greater on roads with a larger proportion of high-income travelers, such as toll roads.

Driver Opinions on WWD Alerts. The next questions were about drivers' opinions toward different methods of notifying right-way drivers of a wrong-way vehicle. Almost all 900 survey respondents (97%) would want to be notified of a nearby WWD event. For the 319 Waze users surveyed, 94% said that WWD alerts should be automatically created in Waze. Of all 900 respondents, most indicated that WEA (63%) or navigation assistants or both (63%) would be their preferred WWD notification method. Figure 2*d* shows that navigational assistance was the preferred WWD notification method for drivers aged 18-50, whereas WEA was preferred by drivers over 50. DMS was preferred by more older drivers than younger drivers. A follow-up question about WEA indicated that 82% of respondents wanted these alerts to be broadcast within a range of 10 mi or less.

Driver Reactions to WWD Alerts. Understanding how drivers would react once they receive a WWD notification is beneficial for operational purposes and to determine if follow-up notifications are needed. Overall, the most popular reaction to a WWD notification would be to pull over and stop on the right shoulder of the road or exit the highway at the nearest exit. Respondents aged 18–35 were more likely to exit the road, whereas those aged 36 and over were more likely pull over to the right shoulder and stop, as seen in Figure 2*e*. This agrees with previous survey findings that showed older drivers were more likely to pull over than younger drivers (*7*). The participants who stated they would pull to the right or left shoulder and stop were asked a follow-up question to determine how long they would remain stopped. As seen in Figure 2*f*, the most common response among all age groups was to wait on the shoulder indefinitely until they receive updated information on the wrong-way driver's location. This indicates that it would be beneficial to distribute follow-up notifications to notify drivers when it is safe to continue their journey.

Waze-Reporting Questions. Since Waze has the potential to be a useful WWD notification system, the 319 respondents who indicated that they were Waze users were asked if and how they would report WWD events in Waze (if the feature was available). Most Waze users (94%) indicated that they would "definitely" or "probably" report a wrong-way driver using the app. Of these respondents, 76% indicated that they would either pull over and report it as soon as possible or would have the passenger exclusively report it. The remaining 24% indicated that they would report it right away while driving. In addition to addressing safety concerns, this information helps determine the time-delay between a driver witnessing and reporting a WWD event, which should be considered for driver-generated alerts. Finally, 61% of Waze users said that they would be more likely to report a WWD event in Waze if a higher point value was assigned to this type of alert. Thus, incentivizing the alert could be beneficial as it could make drivers more likely to report WWD events.

WWD Advanced Countermeasure Questions (All Four Surveys)

The final survey questions involved two ITS WWD advanced countermeasure solutions that have been implemented in Florida. Respondents from all four surveys were asked to compare these two technologies. They were shown two short videos: the first showed one set of "Wrong Way" signs (one sign on the left and one sign on the right) equipped with flashing LEDs around the border of the signs, as implemented by FTE in South Florida. The second video showed two sets of "Wrong Way" signs equipped with RFBs above and below the sign, as implemented by CFX in Central Florida. Both videos were taken from the perspective of a driver traveling the wrong-way up an exit ramp and triggering the devices, and both took place in a similar nighttime setting with vehicles traveling at roughly the same speed. Images from these videos are shown in Figure 3. Respondents were required to watch both videos in their entirety before indicating which alternative they preferred.

Of the combined 2,052 respondents from all four surveys, 74% preferred the RFB signs. The individual results for each survey are shown in Figure 4. After selecting an alternative, respondents of all surveys except the FTE survey were asked why they preferred the alternative they chose. Those who preferred the RFB signs could choose between better flashing, the additional set of signs, or both, whereas those who preferred the LED signs could choose between better flashing, better lighting, or both. Of those who preferred the RFBs, 69% selected "both," whereas 51% of those who preferred the LEDs selected "both."

Finally, an open-ended WWD countermeasure question was given to FHP officers, asking them what solutions they thought were possible to prevent wrong-way driving on limited-access highways. This free response question had a variety of answers, but popular solutions included signs with flashing lights or sirens, tire deflation devices (such as spike strips), more signs, bigger or better lit signs, advanced detection systems (cameras and sensors), and increased law enforcement. Some of these suggestions, such as signs with flashing lights and detection systems, are already in place in Florida and other parts of the United States. Even though many respondents suggested spike strips, a previous TxDOT study found that installing these at highway ramps presents a "significant danger" to the traveling public and recommends against using them as a WWD countermeasure (*24*).

Conclusions and Future Research

During a WWD event, it is critical to provide LEOs and drivers with fast, relevant, and accurate alerts. For LEOs, fast and precise updates about the wrong-way vehicle's location can help them safely and quickly stop the offending vehicle. Agencies have typically relied on DMS to disseminate WWD information to drivers, but these have a limited reach and are often not cost-effective solutions, especially in rural areas. Through surveys conducted on LEOs and drivers, the results from this

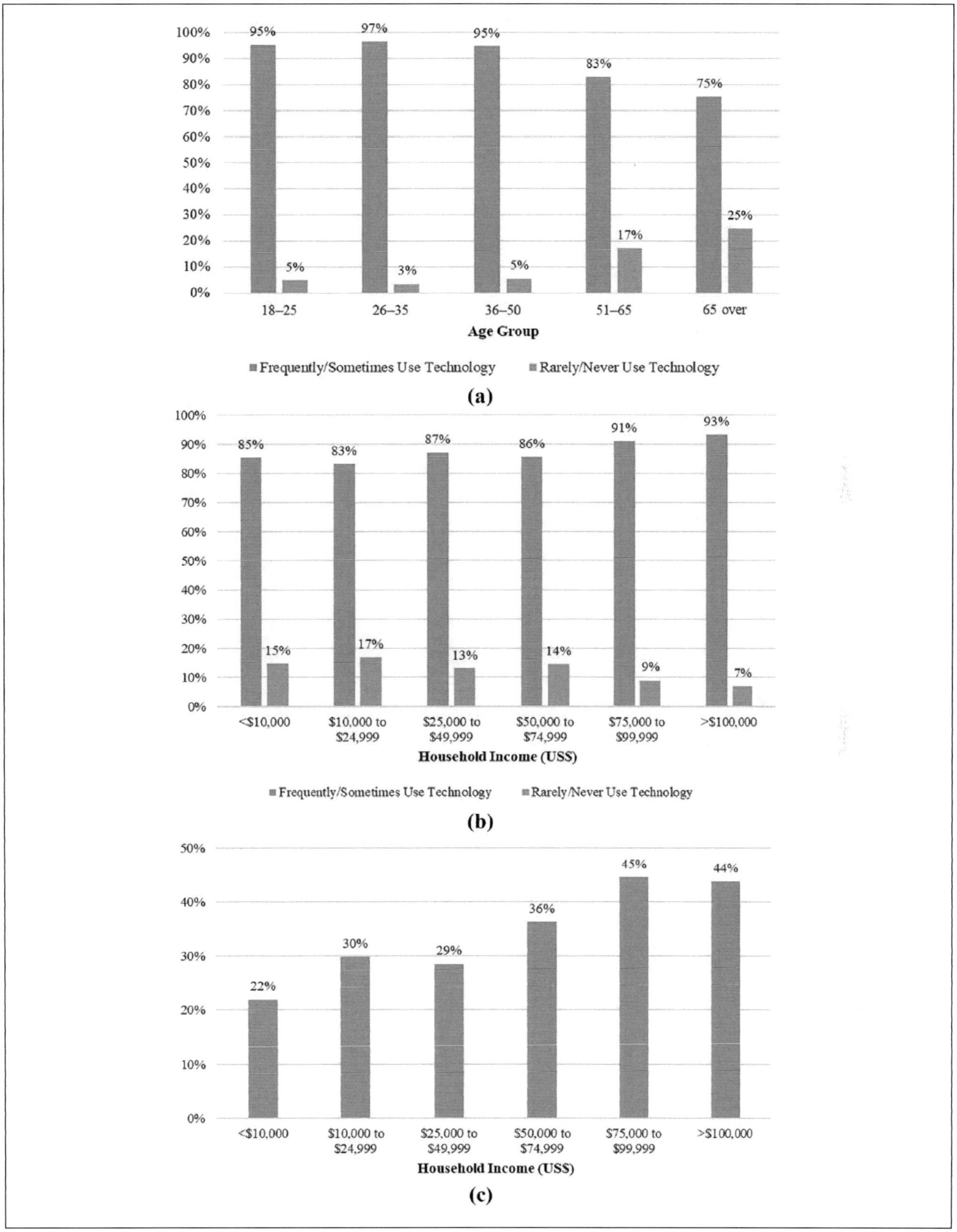

Figure 2. (Continued)

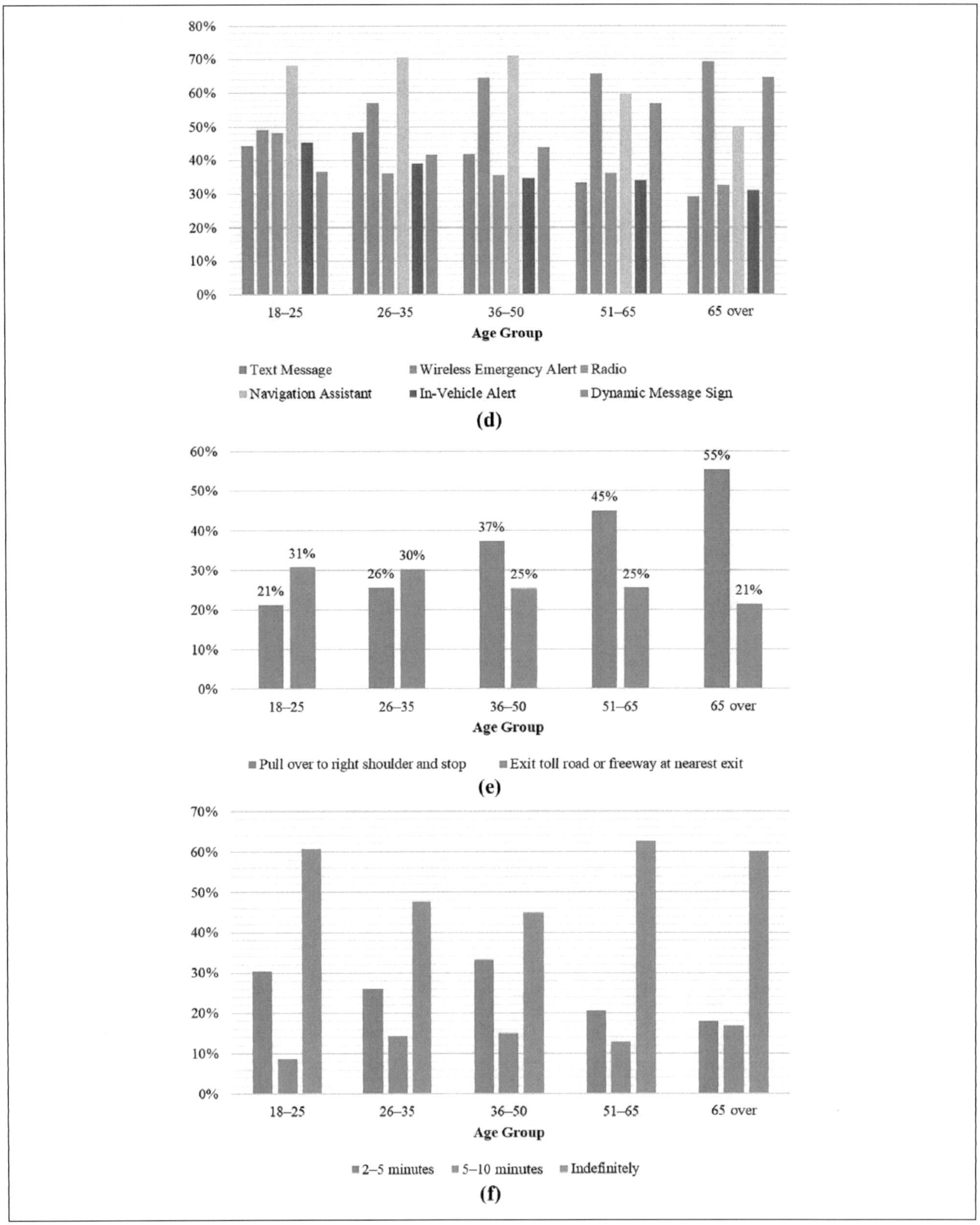

Figure 2. Select response frequency percentages for different age groups and income brackets: (*a*) age group vs. use of navigation assistants; (*b*) household income (US$) vs. use of navigation assistants; (*c*) household income (US$) vs. Waze usage; (*d*) age group vs. WWD notification preference; (e) age group vs. driver reaction; (*f*) age group vs. wait time on shoulder.

Figure 3. Images from WWD countermeasure videos: (*a*) image from video of FTE LED wrong-way signs; (*b*) image from video of CFX RFB double set of wrong-way signs.

research could help agencies implement the most effective solutions to warn right-way drivers and combat WWD.

The law enforcement survey found that most FHP officers are dispatched multiple times per year to WWD events, but typically issue less than one WWD citation per year. The remaining dispatches are either crashes (typically 2 or 3 per officer per year) or drivers that are never found, with most officers stating that over 60% of WWD dispatch events they have worked on have gone on arrival. Additionally, 83% of officers said it is difficult to determine where wrong-way drivers enter the highway. According to the surveyed officers, DUI and confused/impaired drivers are the most common influences associated with nighttime and daytime WWD citation events, respectively. Nevertheless, intoxication remains a high influence in both daytime and nighttime citations, and DUI citations are coupled with 77% of WWD citations. By understanding the characteristics of wrong-way drivers, agencies could implement more effective countermeasures that could result in fewer dispatch events and crashes.

According to the CFX survey, almost all drivers (97%) wanted to be warned of nearby WWD events. Drivers rely mostly on navigation assistants (PNDs and smartphone apps) for travel information, suggesting they would be more likely to receive a notification delivered through these systems. Since most survey respondents prefer warnings transmitted through navigation assistants and WEA, notifications tailored to these methods could effectively reach right-way drivers in their preferred manner. Nearly all Waze users (94%) were receptive to having WWD detection infrastructure automatically generate alerts in Waze.

Reporting WWD events through crowdsourcing services such as Waze could benefit LEOs and other drivers by

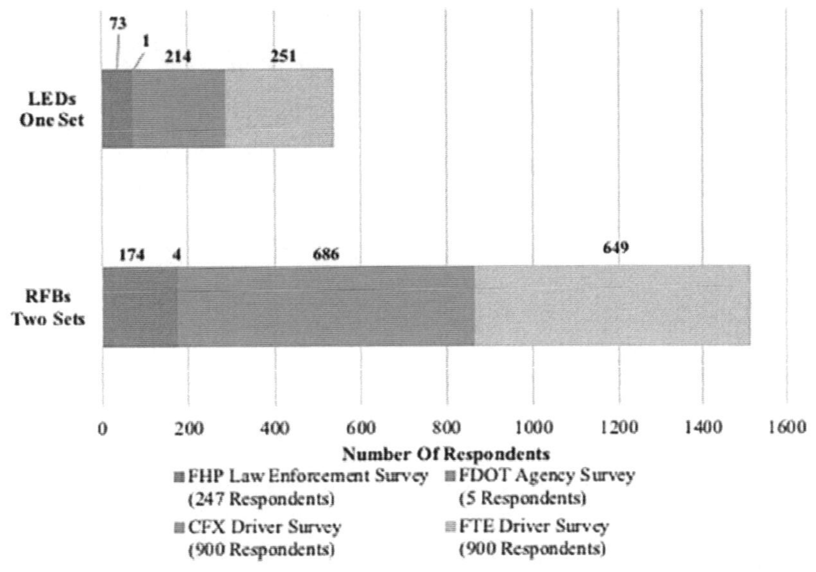

Figure 4. Survey responses on preferred ITS WWD advanced countermeasure.

providing quick and precise updates on a wrong-way vehicle's location, even where there is no ITS WWD detection infrastructure in place. When 319 Waze users were asked about WWD reporting in Waze, 94% indicated that they would "definitely" or "probably" report a wrong-way driver through the app. Most of them (76%) said they would do it in a safe manner by pulling over or exclusively having the passenger report it.

Understanding drivers' reactions to WWD alerts is beneficial for agencies so they can determine when, where, and how WWD event notifications should be administered. According to the survey, most drivers (65%) would either stop on the right shoulder of the road or exit the highway when notified of a WWD event, which are generally the safest options. Older drivers were more likely to pull over than younger drivers. Over half the respondents who stated they would pull over and stop would wait to receive more information on the wrong-way driver's location before continuing their route. This suggests that agencies should broadcast secondary alerts after a WWD event is cleared. Future models on the correlations between area demographics and driver behavior, preferences, and reactions could be developed to provide a comprehensive notification strategy for different areas.

Central Florida's unique WWD countermeasure, which utilizes two sets of RFB signs, proved popular among drivers, LEOs, and FDOT officials, with 70% of LEOs and 74% of drivers preferring these devices over South Florida's single set of LED signs. This was mainly due to both the better flashing and the additional set of signs. Future research is required to perform a full cost-benefit evaluation for these countermeasure systems.

Overall, the results of these surveys could help agencies to determine the safest and most effective ways to notify right-way drivers of WWD events. As most drivers rely on navigation devices and apps, incorporating WWD alerts into these services could allow them to reach more drivers than possible through DMS alone. Agencies could consider WEA for WWD notifications, keeping in mind that most drivers would only want to see these alerts at a range of 10 mi or less. Waze users were overwhelmingly receptive to reporting WWD events through Waze which, if available, could benefit LEOs and other drivers by providing fast and precise updates on the wrong-way driver's location. This study found that it is also important for agencies to notify drivers when a WWD event has been cleared so drivers who pulled over know it is safe to resume driving. By considering drivers' WWD notification preferences, reactions, and current technology use, transportation authorities and law enforcement agencies could improve the safety of both their customers and officers.

Acknowledgments

The research and results discussed in this paper are from projects funded by the Florida Turnpike Enterprise and Central Florida Expressway Authority.

Author Contributions

The authors confirm contribution to the paper as follows: study conception and design: Wertanen, Staves, Al-Deek, Sandt, Carrick, and Rogers; data collection: Wertanen, Staves, Al-Deek, Sandt, Carrick, and Rogers; analysis and interpretation of results: Wertanen, Staves, Al-Deek, Sandt, Carrick, and Rogers; draft manuscript preparation: Wertanen, Staves, Al-Deek, Sandt, Carrick, and Rogers. All authors reviewed the results and approved the final version of the manuscript.

References

1. Wilson, B. Is There a Right Way? *Roads & Bridges*, January 2013. https://www.roadsbridges.com/safety-2013-there-right-way. Accessed July 15, 2017.
2. Sandt, A., H. Al-Deek, J. Rogers, and A. Alomari. Wrong-Way Driving Prevention: Incident Survey Results and Planned Countermeasure Implementation in Florida. *Transportation Research Record: Journal of the Transportation Research Board*, 2015. 2484: 99–109.
3. National Transportation Safety Board. *Wrong-Way Driving*. Highway Special Investigation Report NTSB/SIR-12/01, Washington, D.C., 2012.
4. Michigan Department of Transportation. *Highway Safety White Paper: 2035 MI Transportation Plan, State Long-Range Transportation Plan*. Final Draft, 2012. http://www.michigan.gov/documents/mdot/MDOT_HighwaySafetyWP_385962_7.pdf. Accessed May 25, 2017.
5. Zhou, H., and M. P. Rouholamin. *Guidelines for Reducing Wrong-Way Crashes on Freeways*. Illinois Center for Transportation, Illinois Department of Transportation, 2014. https://apps.ict.illinois.edu/projects/getfile.asp?id=3118. Accessed June 14, 2017.
6. Cooner, S. A., A. S. Cothron, and S. E. Ranft. *Countermeasures for Wrong-Way Movement on Freeways: Overview of Project Activities and Findings*. Publication FHWA/TX-04/4128-1. Texas Transportation Institute, College Station, Texas, 2004.
7. Sandt, A., H. Al-Deek, J. Rogers, A. Alomari, and E. Gordin. Modeling Driver Responses to Wrong-Way Driving Countermeasures Through a Driver Survey and Countermeasure Implementation in Florida. Presented at 95th Annual Meeting of the Transportation Research Board, Washington, D.C., 2016.
8. Chandler, C., D. Buidens, G. Reynolds, A. Young, C. McKenzie, and J. Roso. *FDOT District Seven Wrong Way Driving Countermeasures Update and Lessons Learned*. Florida Department of Transportation, 2017.
9. Venglar, S., and B. Fariello. Efforts to Reduce Wrong-Way Driving: A Case Study in San Antonio, Texas. Presented at 93rd Annual Meeting of the Transportation Research Board, Washington, D.C., 2014.
10. Fox 19 Now. *Wrong Way Crashes on Rise in Tri-State*. September 19, 2016. http://www.fox19.com/story/33130527/wrong-way-crashes-on-the-rise-in-the-tri-state. Accessed July 31, 2017.
11. Oeser, M., T. Volkenhoff, D. Kemper, and C. Wietfeld. Wrong Way Driving on German Motorways: Safety Gain by a Low Cost Detection System. Presented at 94th Annual Meeting of the Transportation Research Board, Washington, D.C., 2015.

12. Simpson, S. *Wrong Way Vehicle Detection: Proof of Concept*. Publication FHWA-AZ-13-697. FHWA, Arizona Department of Transportation, 2013.
13. Rocchio, R. Wrong-Way Driving Mitigation. Rhode Island Department of Transportation. Presented at 94th Annual Meeting of the Transportation Research Board, Washington, D.C., 2015.
14. Al-Deek, H., S. Venkata, and J. Flick. The Impact of Dynamic Message Signs on Satisfaction and Diversion of Toll Road Travelers. *Transportation Research Record: Journal of the Transportation Research Board*, 2009. 2099: 1–13.
15. Cronkite News Arizona PBS. *Seven Wrong-Way Crashes on Arizona Freeways in 2017*. April 14, 2017. https://cronkite-news.azpbs.org/2017/04/14/seven-wrong-way-crashes-on-arizona-freeways-in-2017/. Accessed May 25, 2017.
16. Simpson, S. *Detection and Warning Systems for Wrong-Way Driving*. Publication FHWA-AZ-15-741. FHWA, Arizona Department of Transportation, 2015.
17. *Wireless Emergency Alerts: Customer Guide*. Customer and Governmental Affairs Bureau, Federal Communications Commission. http://transition.fcc.gov/cgb/consumerfacts/wea.pdf. Accessed May 25, 2017.
18. McGregor, J. D., J. P. Elm, and T. Marshall-Keim. *Best Practices in Wireless Emergency Alerts*. Publication CMU/SEI-2013-SR-015, DM-0000430. Software Engineering Institute, US Department of Homeland Security, 2013.
19. *Wireless Emergency Alerts: Arbitrary-Size Location-Aware Targeting*. Science and Technology, US Department of Homeland Security, 2015.
20. Finley, M., K. Balke, R. Rajbhandari, S. Chrysler, C. Dobrovolny, N. Trout, P. Avery, D. Vickers, and C. Mott. *Conceptual Design of a Connected Vehicle Wrong-Way Driving Detection and Management System*. Technical Report FHWA/TX-16/0-6867-1. Texas A&M Transportation Institute, April 2016.
21. Richey, C. Waze Partnership Helping to Reduce Response Times for Emergency Personnel. *Genesis Pulse*, August 16, 2016. https://genesispulse.com/2016/08/16/waze-genesis-partnership/. Accessed July 15, 2017.
22. Malone, K. Blazing the Waze: FDOT Is the Traffic App's First U.S. Partner. *WRLN Public Radio and Television*. http://wlrn.org/post/blazing-waze-fdot-traffic-app-s-first-us-partner. Accessed July 15, 2017.
23. Szoldra, P. People Say the Best Feature of This Hot GPS App Has Nothing to Do with Navigation. *Business Insider*, November 19, 2015. http://www.businessinsider.com/google-waze-gps-app-2015-11. Accessed June 28, 2017.
24. *Engineering Analysis of the Installation of Spike Strips And Other Destructive Devices in Freeway Exit Ramps*. Texas Department of Transportation. http://www.transguide.dot.state.tx.us/sat/wwd/content/EngineeringAnalysisSpikeStrips.pdf. Accessed July 25, 2017.

The Standing Committee on Traffic Law Enforcement (ANB40) peer-reviewed this paper (18-03497).

This paper and its contents, including the conclusions and results, are solely those of the authors; they do not represent the opinions or policies of the funding agencies.

Article

TRR
JOURNAL OF THE TRANSPORTATION RESEARCH BOARD

Transportation Research Record
2018, Vol. 2672(33) 56–66

DOI: 10.1177/0361198118777356
journals.sagepub.com/home/trr
SAGE

Development of Macro-Level Crime and Collision Prediction Models to Support Data-Driven Approach to Crime and Traffic Safety (DDACTS)

Emmanuel A. Takyi[1], Seun Daniel Oluwajana[2], and Peter Y. Park[2]

Abstract

The number of violent crimes and fatal-injury collisions concerns many jurisdictions. Traditional enforcement tactics are often reactive, relying on historical crime and collision data to select locations for law enforcement. Advanced law enforcement tactics take a proactive approach. Such tactics include Data-Driven Approaches to Crime and Traffic Safety (DDACTS), which uses predicted numbers of crimes and collisions to identify locations for law enforcement. This DDACTS study was conducted in Regina, Saskatchewan, Canada. The research developed macro-level prediction models to predict violent crimes and collisions in each traffic analysis zone (TAZ) in Regina. The zonal nature of the analysis is important for overcoming confidentiality and privacy issues associated with violent crimes and fatal-injury collisions. Fifty-four input variables were used to describe each TAZ's crimes, collisions, socio-demographic, road inventory, traffic, and land use characteristics. The analysis used negative binomial regression coupled with the empirical Bayes method (a popular approach in transportation, but relatively new to crime mapping) to develop two statistical models that predict the long-term mean value for the number of violent crimes/collisions per zone. Cumulative residual plots were used as the main goodness-of-fit test. The findings are summarized on a map showing the top ten hotzones for violent crimes, the top ten hotzones for fatal-injury collisions, and the zones where the crime and collisions zones overlap. The overlapping zones are the DDACTS zones. By focusing law enforcement in the DDACTS zones, it may be possible to reduce violent crimes and fatal-injury collisions simultaneously and use limited resources more cost effectively.

The purpose of this study is to develop sophisticated macro-level (zonal-level) crime and collision prediction models to support the advanced place-based law enforcement system known as "Data-Driven Approaches to Crime and Traffic Safety (DDACTS)." Crime and collision data were collected from Regina, Saskatchewan, Canada. Regina is the capital of Saskatchewan and had a population of 247,200 in 2016 (*1*). Various confidentiality and privacy issues are clearly associated with crime data. As it may not be possible to disclose, for example, the specific locations of some crimes, many crime studies have taken a macro-level approach (*2–4*).

This study focuses on violent crimes and fatal-injury collisions, that is, the most serious/severe types of crimes and collisions. By analyzing the locations of recorded crimes and collisions and using the analysis to predict the expected future locations of crimes and collisions (defined as hotzones in this study), the study is expected to enable Regina Police Service to concentrate their law enforcement resources in the way that is most likely to simultaneously and cost effectively reduce the number of crimes and collisions.

Strategies for Law Enforcement

Traditional law enforcement strategies rely heavily on responding to calls and on random or sporadic patrols. These strategies are essentially reactive rather than proactive: police officers typically have to wait until an area is identified as having a crime or traffic collisions problem before taking law enforcement action. This reactive approach to law enforcement is likely to include intuitive

[1]Department of Civil, Geological, and Environmental Engineering, University of Saskatchewan, Saskatoon, SK, Canada
[2]Department of Civil Engineering, Lassonde School of Engineering, York University, Toronto, ON, Canada

Corresponding Author:
Address correspondence to Peter Y. Park: peter.park@lassonde.yorku.ca

and even random elements that may not be cost effective (*5*).

In contrast, proactive enforcement systems aim to predict the occurrence of crimes and collisions and apply appropriate law enforcement measures to reduce the number and/or severity of crimes and collisions, and even to prevent some crimes and collisions from taking place at all (*6*). Proactive enforcement systems include sophisticated intelligence-led policing, predictive policing, selective enforcement, and smart enforcement. Proactive approaches based on prevention through data-supported intelligence are designed to increase cost effectiveness, especially if crimes and collisions can be reduced simultaneously (*7–9*).

Data-Driven Approaches to Crime and Traffic Safety (DDACTS)

DDACTS is a proactive enforcement system used in North America (*10*). Combined with highly visible traffic enforcement, DDACTS is regarded as a sustainable, flexible, and cost-effective method for reducing crime and collisions (*9*). The system seeks to improve the effectiveness and efficiency of law enforcement operations by enabling police officers to conduct focused, highly visible, law enforcement policing tactics based on independent, scientific, and evidence-based crime and collision data analysis. Highly visible traffic enforcement in an area helps to reduce the public's fear of crimes/collisions in that area. It can also have a positive effect on adjacent neighborhoods by reducing crime and collisions in those neighborhoods (the spillover effect). As a DDACTS strategy aims to reallocate the areas targeted for law enforcement rather than change the overall frequency or duration of law enforcement, the budget effect is expected to be neutral: police officers will be assigned to certain locations for law enforcement with or without a DDACTS strategy (*6*).

The use of a DDACTS strategy provides a scientifically justifiable approach to the selection of areas for enhanced law enforcement. This is important for the local community, for local administrators, and for the police force. Under the current approach, some members of the public may, for example, become suspicious of law enforcement and may complain of profiling or revenue-motivated ticketing (*5*). Without a clear justification for the areas targeted for law enforcement, it is difficult to counter these perceptions.

In the United States, DDACTS is recognized as an intelligence-led policing business model that can help police officers to increase the effectiveness of law enforcement in a cost-efficient way. The approach has become a valuable policing strategy. Jurisdictions currently successfully applying DDACTS include many metropolitan areas (e.g., Baltimore County, MD, and Nashville, TN) and many small towns (such as St. Albans, VT, and Sheboygan, WI) (*11,12*). In Baltimore, for example, DDACTS was introduced in early 2009. The number of crimes in the three years before DDACTS was compared with the six-month period (April to November, 2009) after DDACTS was introduced. The reduction in crimes included a 1.1% reduction in thefts from motor vehicles, a 16.6% reduction in burglaries, a 33.5% reduction in robberies, a 40.9% reduction in motor vehicle thefts, and a 43% reduction in injury collisions (*13*).

Place-Based Law Enforcement

Past research has shown that there is a relationship between dangerous driving behavior (e.g., reckless or impaired driving) and criminal behavior. Researchers have developed a number of criminological and psychological studies and theories to explain the overlap between crimes and collisions (*9,14,15*). The studies suggest that individuals who are involved in risky behavior in one activity (e.g., offensive and antisocial activities) will likely be involved in risky behavior in other activities (e.g., excessive speeding or drug-impaired driving). The studies also found that such individuals often exhibit more than one behavioral problem. For example, individuals with a lack of self-control may engage in persistent risk-taking behavior (*9,16*).

Some crimes/collisions do not have identifiable causation motives or factors, but most crimes/collisions have a known place. Crimes and collisions (problems) and law enforcement (fixing problems) therefore have an important geographic element (*17*). Many recent studies in geographic criminology show that some places attract individuals who commit crimes and/or traffic violations (some of the traffic violations presumably lead to collisions). Such places are typically "rough" neighborhoods with high crime rates. They may be considered typical examples of a geographic criminology theory known as deviant place theory (*18*).

From these perspectives, the idea of hotzone areas where crime and collisions overlap becomes very likely, and a statistical analysis using, for example, a DDACTS approach can be expected to show the hotzones. The idea that the hotzones can be predicted also seems likely. It is also reasonable to expect that law enforcement that concentrates on the crime/collision overlap areas will likely reduce crimes and collisions at that location simultaneously.

It is not uncommon for law enforcement agencies to use GIS techniques to create crime maps or collision maps and then inspect the maps for locations showing a high concentration of crimes or collisions. Unfortunately, however, simple visual inspection can easily be misleading, as the determination of genuine hotzones is not straightforward. For example, if a history of crime/collision occurrence numbers is used, the hotzones are likely to move and may change from year to year, complicating the identification of areas where crimes and collisions consistently overlap. This phenomenon occurs as a result of regression-to-the mean bias (*19–21*). Regression-to-the-mean refers to a statistical tendency. In this case, it is important because locations may experience an extreme value for crimes/collisions in one year and then, without any intervention, experience less extreme values in subsequent years. To avoid the pitfalls of regression-to-the-mean bias in crime/collision data analysis, enforcement agencies need maps based on the long-term mean value of crimes/collisions. The creation of these maps requires rigorous spatial data analysis that uses appropriate and advanced statistical modeling techniques.

During the last two decades, there have been intensive discussions regarding the development of better-fitting crime/collision prediction models using a wide range of advanced statistical methods (*22*). The negative binomial regression model coupled with the empirical Bayes (EB) technique is currently regarded as one of the most popularly used methods for developing collision prediction models (*23–27*), but the current standard approach to mapping DDACTS zones uses kernel density analysis based simply on the observed number of crimes and collisions. In this study, the authors use a negative binomial (NB) regression model coupled with the EB technique to develop predict the number of crimes and collisions. This is done to militate against regression-to-the-mean bias, and to the authors' knowledge this is the first DDACTS study to do so. Different models were developed for the number of violent crimes and for the number of fatal-injury collisions.

Study Objectives

The Regina DDACTS study had two main objectives:

- Develop macro-level crime and collision prediction models using an advanced statistical technique; and
- Demonstrate how the Regina Police Service can use the models' results to identify the hotzones where significant numbers of violent crimes and fatal-injury collisions overlap and where law enforcement should be focused (i.e., DDACTS zones).

Study Data

The study used five years (2009 to 2013) of Regina crime, collision, traffic volume, socio-demographic, and land use data to develop the crime and collision prediction models. Traffic analysis zones (TAZs) were used as the spatial analysis unit. There were 299 traffic TAZs in Regina. The data were supplied by three different agencies:

- Five years (2009 to 2013) of crime data (GIS shapefile format) were provided by Regina Police Service;
- Five years (2009 to 2013) of collision data (Microsoft Access format) were supplied by Saskatchewan Government Insurance;
- Five matching years (2009 to 2013) of traffic volume data (GIS shapefile format) were supplied by the City of Regina;
- Roadway inventory data (GIS shapefile format) were supplied by the City of Regina. The data included the location of every intersection and roadway segment in the city (i.e., the roadway network) and details such as the functional classification and speed limit of every roadway segment; and
- Socio-demographic data (Microsoft Excel format) for each TAZ in Regina were supplied by the City of Regina.

During the five-year study period, there were 18,232 violent crimes and 5,759 fatal-injury collisions. The violent crime data refer to five specific categories of crime (assault, arson, robbery, sexual assault, and murder).

Table 1 provides an overview of the study's 54 input variables (two response variables [i.e., violent crimes and fatal-injury collisions] and 52 explanatory variables). All 54 input variables are aggregated to give a single-value for each TAZ. Note that the City of Regina chose and categorized the socio-demographic variables when the City collected data for each TAZ. The City's main purpose was to forecast traffic volumes on their road network. This study used the same data, but recognized that the City of Regina's data choices might not be ideal for the study purpose. For example, the authors might have preferred some additional roadwork variables and different age and education categories.

Methodology

Negative Binomial Regression Model

This study used NB regression in the development of the crimes and collision prediction models. The functional form of the NB model employed in this study is given as Equation 1 for the expected mean (μ_i) which is then re-expressed in Equations 2 and 3 for each zone i:

$$\ln(\mu_i) = x_i^T\beta + \varepsilon_i \quad (1)$$

Table 1. Description of Input Variables

Input variable	Description (per traffic analysis zone)	Mean	Std. Dev	Min	Max
Crime data					
Violent crimes	Total number of violent crimes	69.06	563.37	0.00	9,115.90
Collision data					
Fatal-Injury collisions	Total number of fatal-injury collisions	21.98	21.65	0.00	103.00
Socio-demographic data					
POP_01to17	Population of persons aged 1–17	161.00	179.45	0.00	782.00
POP_18to24	Population of persons aged 18–24	80.00	84.12	0.00	339.00
POP_25to44	Population of persons aged 25–44	215.00	220.58	0.00	884.00
POP_45to64	Population of persons aged 45–64	204.00	214.47	0.00	911.00
POP_65plus	Population of persons aged 65 or above	104.00	125.84	0.00	763.00
TOT_POP	Total population	765.00	785.32	0.00	3,011.00
POP_DENSITY	Population density (persons/km^2)	2,006.34	1,665.67	0.00	10,552.61
NO_GRDSCH	Number of persons enrolled in graduate school	112.00	261.35	0.00	1,720.00
Road inventory characteristics					
AVE_SPDLIM	Weighted average speed limit (km/h)	85.03	55.84	40.00	536.17
TOT_ROADLEN_20	Roadway length with posted speed 20 km/h	0.76	12.33	0.00	200.00
TOT_ROADLEN_30	Roadway length with posted speed 30 km/h	4.50	45.01	0.00	523.00
TOT_ROADLEN_40	Roadway length with posted speed 40 km/h	636.09	818.10	0.00	3,836.75
TOT_ROADLEN_50	Roadway length with posted speed 50 km/h	3,147.48	2,703.98	0.00	14,405.84
TOT_ROADLEN_60	Roadway length with posted speed 60 km/h	28.57	151.33	0.00	1,695.00
TOT_ROADLEN_70	Roadway length with posted speed 70 km/h	122.42	392.76	0.00	3,284.75
TOT_ROADLEN_80	Roadway length with posted speed 80 km/h	130.25	431.64	0.00	3,677.00
TOT_ROADLEN_100	Roadway length with posted speed 100 km/h	70.10	251.46	0.00	1,820.00
TOT_ROADLEN	Total road length, m	4,140.16	3,299.99	87.00	16,354.95
AVE_ROADLEN	Weighted average roadway length, m	2,150.34	1,828.17	81.17	11,361.34
NO_3LEGS_INT	Number of intersections with 3 legs	9.02	10.02	0.00	66.00
NO_4LEGS_INT	Number of all intersections with 4 legs	5.21	6.41	0.00	36.00
NO_5LEGS_INT	Number of all intersections with 5 legs	0.02	0.12	0.00	1.00
TOTAL_INT	Total number of intersections	14.22	13.63	0.00	79.00
INT_DENSITY	Intersection density, (number of intersections per TLKM)	3.19	1.79	0.00	18.35
ARTERIAL_LEN	Arterial length, m	555.93	565.05	0.00	2,618.06
COLLECTOR_LEN	Collector length, m	621.16	751.80	0.00	3,452.70
GRAVEL_ROAD_LEN	Gravel road length (m)	136.50	488.31	0.00	4,879.00
EXPRESSWAY_LEN	Expressway length, m	84.11	262.64	0.00	2,029.00
HIGHWAY_LEN	Highway length, m	41.26	205.32	0.00	1,820.18
LOCAL_ROAD_LEN	Local road length, m	2,369.14	2396.95	0.00	9,643.52
PRIVATE_ROAD_LEN	Private road length, m	235.89	995.59	0.00	12,726.00
RAMP_LEN	Ramp length, m	90.96	275.05	0.00	1,825.64
ROW_LEN	Right of way length, m	4.10	32.73	0.00	434.00
TLKM	The sum of length of all road segments in a TAZ, km	4.14	3.30	0.09	16.35
I3WP	Number of 3-leg intersections as percentage of total number of intersections, %	54.38	34.22	0.00	100.00

(continued)

Table 1. (Continued)

Input variable	Description (per traffic analysis zone)	Mean	Std. Dev	Min	Max
ALKP	Arterial road length as percentage of total-lane-kilometers, %	19.82	23.85	0.00	100.00
LLKP	Local road length as percentage of total-lane-kilometers, %	47.38	30.64	0.00	100.00
Traffic volume					
VKT	Vehicle-kilometer traveled	23,899.17	21,490.66	43.43	122,800.88
Land Use					
COMMERCIAL_AREA	Total commercial area, m^2	28,150.48	46,800.05	0.00	312,623.00
INSTITUTIONAL_AREA	Total institutional area, m^2	16,050.20	46,778.39	0.00	620,487.35
OPENSPACE_RECREATION_AREA	Total area designated for open space/recreation purposes, m^2	93,818.46	248,948.94	0.00	2,501,580.25
RAILWAY_AREA	Total area designated for railway use, m^2	10,560.52	52,694.82	0.00	691,835.00
RESIDENTIAL_HD_AREA	Total high density residential area, m^2	16,016.33	32,603.35	0.00	196,674.39
RESIDENTIAL_LD_AREA	Total low density residential area, m^2	172,483.88	210,194.97	0.00	786,855.00
RESIDENTIAL_MD_AREA	Total medium density residential area, m^2	13,932.13	43,650.73	0.00	383,555.00
URBAN_HOLDING_AREA	Area intended to protect lands required for future urban developments, m^2	83,510.43	322,005.57	0.00	3,150,675.91
OFFICE_AREA	Office floor space area (m^2)	3,431.05	10,266.25	0.00	101,970.00
RETAIL_AREA	Retail floor space area (m^2)	6,192.72	13,029.34	0.00	122,199.00
INDUSTRY_AREA	Industry floor space area (m^2)	6,134.17	24,540.63	0.00	274,205.00
HOSPT_SPACE	Hospital floor space (m^2)	445.75	4,732.10	0.00	66,704.00
URBAN_HOLDING_RESIDENTIAL_AREA_PROP	Proportion of TAZ area designated as urban holding residential area	0.17	0.34	0	1.00
NO_LU_PER_TAZ	Numbers of different land uses per traffic analysis Zone	4.13	1.68	1.00	8.00
TAZ_AREA	Area of traffic analysis zone, m^2	521,414.91	570,457.13	0.00	6,044,326.67

$$\mu_i = exp\left(x_i^T \beta + \varepsilon_i\right) \quad (2)$$

$$y_i \sim x_i^T \beta \quad (3)$$

where, y_i is the observed value of the response variable (i.e., the number of violent crimes or fatal-injury collisions) for zone i,

x_i is the explanatory variable for zone i, and;

β is the coefficient for zone i.

y_i follows a NB distribution. The probability density function for the NB model can be written as Equation 4 with dispersion parameter α.

$$P(y_i|\mu_{i,}\alpha) = \frac{\Gamma\left(\frac{1}{\alpha} + y_i\right)}{\Gamma(y_i + 1)\Gamma\left(\frac{1}{\alpha}\right)} \left(\frac{1}{1 + \alpha\mu_i}\right)^{\frac{1}{\alpha}} \left(\frac{\alpha\mu_i}{1 + \alpha\mu_i}\right)^{y_i} \quad (4)$$

The likelihood function of Equation 4 is:

$$L(\mu|y, \alpha) = \prod_{i=1}^{n} \exp\left\{y_i \ln\left(\frac{\alpha\mu_i}{1 + \alpha\mu_i}\right) - \frac{1}{\alpha}\ln(1 + \alpha\mu_i) + \ln\Gamma\left(\frac{1}{\alpha} + y_i\right) - \ln\Gamma(y_i + 1) - \ln\Gamma\left(\frac{1}{\alpha}\right)\right\} \quad (5)$$

The log likelihood function of the NB model is shown in Equation 6:

$$\ln(\beta|y, \alpha) = \sum_{i=1}^{n} y_i ln\left(\frac{\alpha exp\left(x_i^T\beta\right)}{1 + \alpha exp\left(x_i^T\beta\right)}\right) - \frac{1}{\alpha} ln\left(1 + \alpha exp\left(x_i^T\beta\right)\right) + ln\Gamma\left(\frac{1}{\alpha} + y_i\right) - ln\Gamma(y_i + 1) - ln\Gamma\left(\frac{1}{\alpha}\right) \quad (6)$$

The study used R-Language (*28*) to estimate the parameters of the NB models developed.

Empirical Bayes Method

Regression-to-the-mean bias is a common issue in collision prediction modeling. To overcome this bias, the EB method was used (*29–31*).

The EB method in Equation 7 uses information (i.e., NB model outputs) from similar zones to estimate the long-term mean values (i.e., the EB-adjusted values) of the number of crimes/collisions in a certain zone (*26,32*):

$$E[y] = w \cdot \mu_i + (1 - w)y_i \quad (7)$$

where, $E[y]$ represents the EB-adjusted value of violent crimes or fatal-injury collisions,

μ_i represents the predicted value of violent crimes or fatal-injury collisions,

y_i represents the observed value of violent crimes or fatal-injury collisions, and

w is the EB weight factor which is given in Equation 8:

$$w = \frac{1}{1 + \alpha \times \sum_{t=1}^{Y} \mu_{ti}} \quad (8)$$

In Equation 8:

α is the dispersion parameter,

μ_{ti} is the predicted number of violent crimes or fatal-injury collisions for zone i in year t, and

Y is the 5 years in this study.

Results of the Analysis

Negative Binomial Regression Model

The analysis was started by developing a correlation matrix for the 54 input variables (the matrix is not presented in this paper because of its large size). The correlation matrix helped to highlight the strong relationships between the response variables and the explanatory variables. Explanatory variables with strong relationships provide potentially strong candidates for producing better-fitting results. For instance, POPULATION_25TO44 had the strongest linear relationship ($r = 0.48$) with crimes, and Vehicle-Kilometer Traveled had the strongest linear relationship ($r = 0.88$) with collisions. It was expected that these two variables to contribute to improving the fitting performance of the crime and collision prediction models respectively.

Using the correlation matrix, the explanatory variables for the crime and collision prediction models were selected following a forward variable selection procedure, that is, the authors added one variable at a time and checked whether the additional variable helped to improve the fitting performance. The most common confidence level (95%) was not used when selecting statistically significant variables for the crime model or the collision model. Hauer has argued that although a variable may not be statistically significant at the 95% confidence level, the variable may still improve the fit and, presumably, the model's predictive performance (*33*). As a result, a variable was added to the crime and collision models if the variable contributed to improving predictive performance.

Various goodness-of-fit (GOF) tests are available to compare the predictive performance of candidate models (*34,35*). The tests include Mean Squared Error, Mean Squared Prediction Error, Mean Prediction Bias, and Mean Absolute Deviation. Residuals are used as an input value for examining a model's fitting performance for all the GOF tests listed. Hauer emphasized the importance of applying cumulative residual (CURE) plots as a main GOF test (*33*). CURE plots show the

Table 2. Models Developed for Violent Crime and Fatal-Injury Collision Prediction

Input variable	Description (per traffic analysis zone)	Estimate	Pr>|Z|
Violent crimes			
Intercept		−0.189	0.195
POP_18to24	Population of persons aged 18–24	0.0048	0.1385
POP_25to44	Population of persons aged 25–44	0.0085	<0.001
POP_45to64	Population of persons aged 45–64	−0.0076	<0.001
POP_DENSITY	Population density (Persons/km^2)	0.0002	0.0022
Ln(COMMERCIAL_AREA)	Total commercial area, m^2 (Log-transformed)	0.574	<0.001
RETAIL_AREA	Retail floor space area, m^2	0.112	0.0799
Dispersion (α)		0.9921	
Fatal-Injury collisions			
Intercept		−3.73	< 0.001
TOT_ROADLEN_70	Roadway length with posted speed 70 km/h	0.0002	0.084
TOT_ROADLEN_80	Roadway length with posted speed 80 km/h	−0.0004	0.0157
NO_3LEGS_INT	Number of intersections with 3 legs	−0.014	0.0317
NO_4LEGS_INT	Number of intersections with 4 legs	0.0304	< 0.001
INT_DENSITY	Intersection density, (number of intersections per TLKM)	0.671	0.0017
Ln(COLLECTOR_LENGTH)	Collector length, m (log-transformed)	−2.6	0.0039
Ln(VKT)	Vehicle-kilometer traveled	0.515	< 0.001
URBAN_HOLDING_ RESIDENTIAL_AREA_PROP	Proportion of TAZ area designated as urban holding residential area	−2.72	< 0.001
Dispersion (α)		0.4641	

magnitude of the residuals and their standard errors along the range of important explanatory variables (e.g., POPULATION_25TO44 and Vehicle-Kilometers Traveled in this study). This study used CURE plots as the main tool for understanding the predictive performance of the crime and collision prediction models developed.

Table 2 shows the parameter values of the explanatory variables used in the two models and the associated *p*-values (i.e., Pr>|Z|). The prediction model for violent crimes included six input variables: COMMERCIAL_AREA, RETAIL_SPACE, and four population-related variables. Commercial and retail areas are associated with business activity, retail stores, and shopping malls, and with the related business and shopping trips. Such locations are often targets for crimes such as robbery and assault, and show a positive association with violent crime. Increased population density also appeared to be associated with an increase in violent crime. The age of the population also appeared to play a role.

A zone with a relatively large population between the ages of 18 and 44 showed a more positive geographic association with violent crime than a zone with a relatively large population in the 45–64 age group. (Note, however, that this finding does not necessarily mean that it was the younger people who were committing the crimes in the zones affected.)

Four of the six input variables had a statistically significant relationship with violent crime at the 95% confidence level. The relationships between POPULATION_18TO24 and violent crime, and RETAIL_SPACE and violent crime were not statistically significant at the same confidence level.

The prediction model for fatal-injury collisions prediction model included eight variables including variables related to speed and intersections. Seven of the eight variables were statistically significant at the 95% confidence level. The exception was road segments with a posted speed limit of 70 km/h.

The results show some interesting relationships. Zones with longer vehicle-kilometers traveled had a higher number of fatal-injury collisions. Road segments with a posted speed limit of 70 km/h, 4-leg intersections and intersection density are all positively associated with fatal-injury collisions. In Regina, roads with a 70 km/h posted speed limit are usually minor arterials with many access points to shopping malls or retail stores, but no center medians. These features of the road may increase the number of fatal-injury collisions. Road segments with a posted speed limit of 80 km/h have a negative association with fatal-injury collisions. This result may appear unexpected, but in Regina, roads with an 80 km/h posted speed limit are usually major arterials with center medians. The center medians likely help to reduce the number of head-on collisions and possibly contribute to reducing fatal-injury collisions. Four-leg intersections can be assumed to have higher traffic volumes and more conflicts than 3-leg intersections, leading to an increase in fatal-injury collisions. Increased intersection density for a zone also contributes to an increase in fatal-injury collisions.

Figure 1 shows the CURE plots (*36*) for the violent crime and fatal-injury prediction models. POPULA

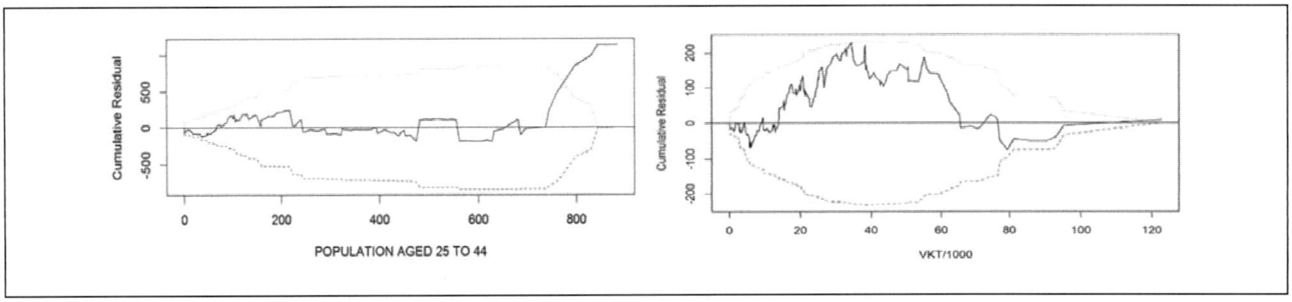

Figure 1. CURE plots for violent crimes (left) and fatal-injury collisions (right).

TION_25TO44 is used in the violent crimes CURE plot, and Vehicle-Kilometers Traveled is used in the fatal-injury collisions CURE plot. The purpose of the CURE plots is to understand the fitting performance of the models. A good CURE plot should move around the zero coordinate of the plot and should tend to converge at zero.

The results show that the two models are generally a good fit. In the case of the violent crime prediction model, however, the upward movement of the cumulative residuals on the right side shows that the model tends to underestimate the predicted values for violent crimes in zones with approximately 750 or more people aged between 25 and 44. The violent crimes models should therefore be used with caution for these particular zones. Omitted variable bias may account for the underestimation in the crime prediction model. It is expected, however, that the underestimation will be reduced when the EB method is used. This is because the predicted value will then be combined with the (probably higher) number of observed violent crimes in the zones.

Crime Hotzones, Collision Hotzones, and DDACTS Zones

The outputs listed in Table 2 were used as inputs to calculate the EB-adjusted values of crimes and collisions (see Equation 7). The calculation was repeated for each zone. Using the value produced, the top ten hotzones for violent crimes and the top ten hotzones for fatal-injury collisions could be listed.

Figure 2 shows the top ten hotzones for violent crimes (blue) and fatal-injury collisions (red). The zones shown in purple are the zones where the violent crimes and fatal-injury collisions hotzones show a significant overlap. These are the four DDACTS zones.

Table 3 presents selected statistics for the top ten hotzones for violent crime and fatal-injury collision and for the four DDACTS zones. For instance, the top ten violent crime hotzones account for 4.5% of Regina's area, but include 35.9% (3,278) of the city's violent crimes for the study period. The top ten fatal-injury collision hotzones account for 5.2% of the city's area, but include 15.5% (892) of the fatal-injury collisions.

The DDACTS zones account for only 1.5% of Regina's area, but include 20.9% of the violent crimes and 6.1% of the fatal-injury collisions. These areas present a suitable focus for Regina Police Service's law enforcement and an opportunity to reduce violent crimes and fatal-injury collisions simultaneously.

Summary and Recommendations for Future Work

This research started with two objectives: developing macro-level crime and collision prediction models; and demonstrating how the models' results can be used to identify the hotzones where significant numbers of violent crimes and fatal-injury collisions overlap. To meet these two objectives, five years (2009–2013) of macro-level data for Regina, Canada were collected, and 52 explanatory variables that described the socio-demographic, road inventory, traffic volume, and land use characteristics for each zone of the study area were considered. Also collected were violent crime and fatal-injury data for each zone. The number of violent crimes and the number of fatal-injury collisions were the two response variables.

A correlation analysis was conducted to pre-screen potentially strong explanatory variables for the violent crime and collision prediction models. NB regression modeling (an industry-standard statistical method) was then used to search for the best-fitting models for crime and collision prediction. Various GOF tests were applied to assess the models, but as the residuals are used as an input for all the GOF tests applied, only the CURE plot results were reported. The CURE plots conveniently show the changes in magnitude of the residuals along the range of values of the selected input variables.

The current standard approach to mapping DDACTS zones uses kernel density analysis based on the observed number of crimes and collisions. This study, however,

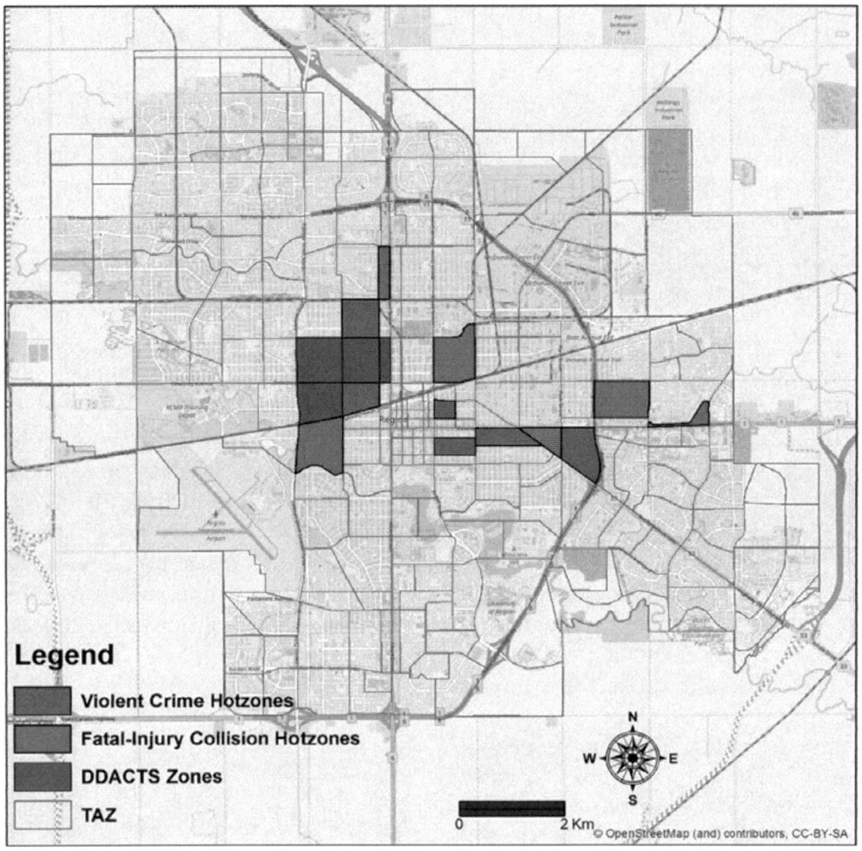

Figure 2. Crime hotzones, collision hotzones, and DDACTS zones in Regina.

Table 3. Selected Statistics for the Top Ten Hotzones and DDACTS Zones

Categories	Violent crimes	Fatal-injury collisions
a. Top ten zones' number of violent crimes or fatal-injury collisions during the study period	3,278	892
b. DDACTS zones' number of violent crimes or fatal-injury collisions during the study period	1,906	352
c. All zones' number of violent crimes or fatal-injury collisions during the study period	9,135	5,758
d. "a" as a percentage of "c"	35.9%	15.5%
e. "b" as a percentage of "c"	20.9%	6.1%
f. Area of top ten zones	4.5 km^2	5.2 km^2
g. Area of DDACTS zones	2.2 km^2	2.2 km^2
h. Area of all zones	147.9 km^2	147.9 km^2
i. "f" as a percentage of "h"	3.0%	3.5%
j. "g" as a percentage of "h"	1.5%	1.5%

used NB regression modeling to predict the number of crimes and collisions and improve the prediction performance of the NB models by using the EB method to produce the long-term mean value of the number of crimes and collisions as the models' outcome. The authors believe this study is the first to try to reduce regression-to-the-mean bias in the number of crimes and collisions concurrently and thereby improve the selection of DDACTS zones for law enforcement.

The study results are presented on a map showing the top ten hotzones for predicted violent crimes and fatal-injury collisions. The map also shows the four DDACTS zones where the violent crimes and fatal-injury collisions overlap and where it may be possible to focus highly visible traffic enforcement to reduce violent crimes and fatal-injury collisions simultaneously. This would enable the police service to increase the cost effectiveness of their enforcement.

Future work may include the development of prediction models and hotzone maps that target other types of crime (e.g., auto theft or theft from auto) and other types of collisions (e.g., property damage only collisions). Additional information of this nature is likely to make further contributions to the cost effectiveness of enforcement.

Future work may include other important factors. For example, weather conditions may have a significant influence on the occurrence of crimes and collisions. Analysis of complex factors such as weather conditions would require additional suitable explanatory variables, some of which would need to be related to certain periods (e.g., seasonal traffic volume, number of shoppers at different seasons of the year, etc.). It may be relatively straightforward to relate weather conditions to existing crime and collision data, but developing and obtaining potentially important explanatory variables that vary with weather conditions is likely to be challenging.

Future work may also include a rigorous before and after study that evaluates how the adoption of DDACTS hotzones for increased police attention affects the number of crimes and collisions and their seriousness/ severity.

Acknowledgments

This research was funded by Saskatchewan Government Insurance (SGI), Regina Police Service, Saskatchewan Centre of Excellence for Transportation and Infrastructure (SCETI), the Natural Sciences and Engineering Research Council of Canada (NSERC), and the City of Regina. The inputs from Dr. Tayyab Shah in this research are greatly appreciated.

References

1. Statistics Canada, Population of Census Metropolitan Area. http://www.statcan.gc.ca/tables-tableaux/sum-som/l01/cst01/demo05a-eng.htm. Accessed July 31, 2017.
2. Andresen, M. A. Crime Measures and the Spatial Analysis of Criminal Activity. *British Journal of Criminology*, Vol. 46, No. 2, 2005, pp. 258–285.
3. Carter, J. G., and E. L. Piza. Spatiotemporal Convergence of Crime and Vehicle Crash Hotspots: Additional Consideration for Policing Places. *Crime & Delinquency*, 2017, pp. 1–25. DOI: 0011128717714793
4. Metz, N., and M. Burdina. Neighbourhood Income Inequality and Property Crime. *Urban Studies*, Vol. 55, No. 1, 2016, pp. 133–150. DOI: https://doi.org/10.1177/0042098016643914
5. Carrick, G., and S. Washburn. Enforcement Engineering: Conceptual Exploration for New Engineering Discipline. *Transportation Research Record: Journal of the Transportation Research Board*, 2011. 2265: 146–152.
6. Silverii, S. Changing the Culture and History of Policy. http://lawenforcementtoday.com/2012/10/09/changing-the-culture-and-history-of-policing/. Accessed March 18, 2013.
7. Hardy, E. Data-Driven Policing: How Geographic Analysis Can Reduce Social Harm. *Geography Public Safety*, Vol. 2, No. 3, 2010, pp. 1–2.
8. Kuo, P. F., D. Lord, and T. D. Walden. Using Geographical Information Systems to Organize Police Patrol Routes Effectively by Grouping Hotspots of Crash and Crime Data. *Journal of Transport Geography*, Vol. 30, 2013, pp. 138–148.
9. Wilson, R. E. Place as the Focal Point: Developing a Theory for the DDACTS Model. *Geography and Public Safety*, Vol. 2, No. 3, 2010, pp. 3–4.
10. National Highway Traffic Safety Administration. *Data-Driven Approaches to Crime and Traffic Safety (DDACTS)*. Operational Guidelines, DOT HS 811 185, National Highway Traffic Safety Administration, Washington, D.C., 2009.
11. Stuster, J., R. Worden, S. Mclean, and J. D. Stuster. *Data-Driven Approaches to Crime and Traffic Safety (DDACTS): Case Studies of Six Programs*. Final Report, Enforcement and Justice Service Division, National Highway Traffic Safety Administration, Washington, D.C., 2010.
12. Wisconsin Department of Transportation. DDACTS: Targeting Crash and Crime Hotspots. *Wisconsin Traffic Safety Reporter*, Vol. 13, No. 3, 2010, pp. 1–3 & 7.
13. Anacapa Sciences, Inc. *Data-Driven Approaches to Crime and Traffic Safety (DDACTS): Case Study of the Baltimore County Police Departments DDACTS Program*. National Highway Traffic Safety Administration, Enforcement and Justice Service Division, California, 2009.
14. Castillo-Manzano, J. I., M. Castro-Nuño, and X. Fageda. Are Traffic Violators Criminals? Searching for Answers in the Experiences of European countries. *Transport Policy*, Vol. 38, 2015, pp. 86–94.
15. Fleiter, J., A. Watson, B. Watson, and V. Siskind. Criminal Histories of Crash and Non-Crash Involved Queensland Speeding Offenders: Evidence Supporting the Idea that We Drive as We Live. *Proc., 2015 Australasian Road Safety Conference*, October 14–16, Gold Coast, Australia, 2015.
16. Brace, C., M. Whelan, B. Clark, and J. Oxley. The Relationship between Crime and Road Safety. Report No. 284. Monash University Accident Research Centre, Clayton, Vic., 2009.
17. Smith, S., and C. Bruce. *CrimeStat III, User Workbook*. National Institute of Justice, Washington, D.C., 2008.
18. Siegel, L. J., and C. R. McCormick. *Criminology in Canada: Theories, Patterns, and Typologies*. Nelson Education Limited, Toronto, 2009.
19. Barnett, A. G., J. C. van der Pols, and A. J. Dobson. Regression to the Mean: What it is and How to Deal with it. *International Journal of Epidemiology*, Vol. 34, No. 1, 2004, pp. 215–220.

20. Senn, S. Francis Galton and Regression to the Mean. *Significance*, Vol. 8, No. 3, 2011, pp. 124–126.
21. Shephard, R. J. Regression to the mean. *Sports Medicine*, Vol. 33, No. 8, 2003, pp. 575–584.
22. Lord, D., and F. Mannering. The Statistical Analysis of Crash-Frequency Data: A Review and Assessment of Methodological Alternatives. *Transportation Research Part A: Policy and Practice*, Vol. 44, No. 5, 2010, pp. 291–305.
23. Hauer, E. Empirical Bayes Approach to the Estimation of "Unsafety": The Multivariate Regression Method. *Accident Analysis and Prevention*, Vol. 24, No. 5, 1992, pp. 457–477.
24. Hauer, E., D. Harwood, F. Council, and M. Griffith. Estimating Safety by the Empirical Bayes Method: A Tutorial. *Transportation Research Record: Journal of the Transportation Research Board*, 2002. 1784: 126–131.
25. AASHTO. *Highway Safety Manual*. AASHTO, Washington, D.C., Vol. 2. 2010.
26. Montella, A. A. Comparative Analysis of Hotspot Identification Methods. *Accident Analysis and Prevention*. Vol. 42, No. 2, 2010, pp. 571–581.
27. Zou, Y., D. Lord, Y. Zhang, and Y. Peng. Comparison of Sichel and Negative Binomial Models in Estimating Empirical Bayes Estimates. *Transportation Research Record: Journal of the Transportation Research Board*, 2013. 2392: 11–21.
28. R Core Team. *R: A Language and Environment for Statistical Computing*. R Foundation for Statistical Computing, Vienna, Austria, 2016. https://www.R-project.org/.
29. Persaud, B., and C. Lyon. Empirical Bayes Before–After Safety Studies: Lessons Learned from Two Decades of Experience and Future Directions. *Accident Analysis and Prevention*, Vol. 39, No. 3, 2007, pp. 546–555
30. Sun, X., and N. J. Garber. *Determining the Safety Effects of Differential Speed Limits on Rural Interstate Highways Using Empirical Bayes Method*. Research Report No. UVACTS-14-5-36. Center for Transportation Studies at the University of Virginia, 2002.
31. Yu, H., P. Liu, J. Chen, and H. Wang. Comparative analysis of the spatial analysis methods for hotspot identification. *Accident Analysis & Prevention*, Vol. 66, 2014, pp. 80–88.
32. Lu, J., K. Haleem, P. Alluri, and A. Gan. Full Versus Simple Safety Performance Functions: Comparison Based on Urban Four-Lane Freeway Interchange Influence Areas in Florida. *Transportation Research Record: Journal of the Transportation Research Board*, 2013. 2398: pp. 83–92.
33. Hauer, E. *The Art of Regression Modeling in Road Safety*. Springer, New York, 2015.
34. Oh, J., C. Lyon, S. Washington, B. Persaud, and J. Bared. Validation of FHWA Crash Models for Rural Intersections: Lessons Learned. *Transportation Research Record: Journal of the Transportation Research Board*, 2003. 1840: 41–49.
35. Washington, S. P., B. N. Persaud, C. Lyon, and J. Oh. *Validation of Accident Models for Intersections*. Publication FHWA-RD-03-037. U.S. Department of Transportation, Washington, D.C., 2005.
36. Hauer, E., and J. Bamfo. Two Tools for Finding What Function Links the Dependent Variable to the Explanatory Variables. *Proc., ICTCT 97 Conference*, Lund, Sweden, 1997.

The Standing Committee on Traffic Law Enforcement (ANB40) peer-reviewed this paper (18-03505).

The findings, opinions and suggestions given in this paper do not necessarily reflect the official perspectives of any of the agencies listed above.

Article

TRR
JOURNAL OF THE TRANSPORTATION RESEARCH BOARD

Transportation Research Record
2018, Vol. 2672(33) 67–77

DOI: 10.1177/0361198118790131
journals.sagepub.com/home/trr
SAGE

Estimating the Effect of Standard Enforcement of a Rear Seat Belt Law for Rear Seat Fatality Prevention

Daniel J. Findley[1], Morgan Sanchez[1], and Timothy Nye[1]

Abstract

Data were collected from the NHTSA's Fatality Analysis Reporting System to analyze the link between primary enforcement of rear seat belt use and injury severity in fatal vehicle collisions. Specifically, this study predicted the amount of fatalities that may have been prevented had there been standard enforcement of a rear seat belt law in effect. Previous literature concludes that increasing seat belt use will decrease injury severity in collisions and the primary enforcement of seat belts laws will increase seat belt use by approximately 14%. This study recorded and compared the number of rear seat fatalities in states that did and did not have primary enforcement laws for rear seat occupants. The results indicated that, on average, for every properly restrained rear seat fatality there are 0.45 more rear seat fatalities in states without primary enforcement than states with primary enforcement. It also predicts that the states that do not practice standard seat belt enforcement could have seen approximately 772 to 1,032 fatalities prevented from 2011 to 2015, had there been primary rear seat belt enforcement. This corresponds to an estimated national crash cost savings of $8.6 billion, or $1.7 billion annually.

The use of seat belts is an effective way to reduce injury severity in motor vehicle collisions. For decades, non-profit and government organizations have tried to increase seat belt usage with various levels of success. For-profit companies, including insurance and health-care providers, have also attempted to increase seat belt use through various methods, including public safety awareness campaigns. One of the most prominent public safety campaigns centered on increasing seat belt use is the *Click It or Ticket* program. This campaign started in North Carolina in 1993 as part of the passing of a primary enforcement seat belt law (*1*). Standard enforcement seat belt laws, also called primary enforcement seat belt laws, allow law enforcement officers to give motor vehicle occupants fines and citations without observing another offense. There are only 18 states that practice standard enforcement of seat belts for rear seat occupants; almost all of the remaining states only have primary enforcement laws for front seat occupants and practice secondary enforcement for rear seat occupants. Secondary enforcement of rear seat belt laws means that vehicles can only be stopped and have rear seat belt citations issued after observing other primary offenses. In the states with secondary enforcement of rear seat belts, law enforcement officers cannot stop a vehicle and issue citations if the only observed infraction is a rear seat passenger not properly restrained.

From 2011 to 2015, 8,285 rear seat passengers in United States died in fatal collisions – 5,069 of these, 61.1%, were unrestrained. Of the states that had a primary law in effect during this time, only 57.2% of their rear seat fatalities were unrestrained, compared with 64.1% for states without a primary law. The NHTSA reported that in states that enact a standard enforcement rear seat belt law, rear seat belt usage increases by approximately 14% (*2*). The increase in seat belt usage by rear seat passengers should result in a prevention of some fatalities and a reduction in injury severity in the more serious and violent collisions, as indicated by the results of previous studies. This reduction in rear seat fatalities and injury severity may also have a significant economic impact by reducing total crash costs.

The primary objectives of this study were to collect data from the NHTSA's Fatality Analysis Reporting System (FARS) and determine the following information:

- The number of fatalities of rear seat occupants in fatal collisions from 2011 to 2015, restrained and unrestrained.

[1]Institute for Transportation Research and Education, Raleigh, NC

Corresponding Author:
Address correspondence to Timothy Nye: tsnye@ncsu.edu

- A prediction or estimate for the number of fatalities that could have been prevented had every rear seat passenger in states without standard enforcement been using proper seat belts or restraints.
- A prediction or estimate for the expected number of fatalities, given likely increases in seat belt use, that could have been prevented if a state without standard enforcement enacted a primary rear seat belt law.

Three methods are presented and considered in this paper to estimate the number of fatalities that may have been prevented with the practice of standard seat belt enforcement:

1. Estimating the number of fatalities prevented had there been 100% usage of seat belts, while considering that the forces in some collisions are too violent for even the properly restrained to survive.
2. Estimating the number of fatalities prevented by comparing the ratio of unrestrained to restrained rear seat fatalities between states with and without standard enforcement.
3. Estimating the number of fatalities prevented had there been a 14% increase in rear seat belt use.

Literature Review

The importance of proper seat belt use and its relationship to injury severity in vehicle collisions has garnered the attention of researchers for decades. When completing a literature review for the purpose of this study, two categories of research were reviewed: (1) studies that compare proper seat belt use and injury severity and (2) studies that examine the effects of seat belt laws on seat belt use.

Proper Seat Belt Use and Injury Severity Comparisons

An estimated 300,000 people die and 10 to 15 million people are seriously injured in traffic crashes throughout the world per year (*3*). Eighty percent of patients requiring hospitalization did not use seat belts, even though data show that belts reduce the incidence of traumatic lesions during motor vehicle crashes (*3*). A 2014 study found that there was a 45% to 60% reduction in fatal injuries for front seat occupants if proper restraints were used, compared with those that were unrestrained (*4*). Data from the United States suggest that the relative risk of death among belted compared with unbelted occupants was approximately 39% (*5*). Of those who died in collisions, 50% of the fatalities could have been prevented with the use of seat belts (*5*). Lap belts alone can reduce the risk of fatalities by 17% to 26% (*6*). A meta-analysis of six studies since 2000 was conducted about increasing seat belt use, with results indicating that seat belts reduce both fatal and non-fatal injuries by 60% among front seat occupants and by 44% among rear seat occupants, with some exceptions because of drunk driving or speeding (*7*). The effectiveness of seat belts in reducing fatal and severe injuries among rear seat passengers is estimated between 32% and 45% based on traffic crash data from countries like the U.S., UK, and Japan (*8*).

The use of seat belts among rear seat passengers does not only have an effect on those in rear seats, but it has also been shown to result in a reduction in injury severity among belted front seat occupants. This is primarily because unrestrained rear seat occupants can become a projectile in the event of a collision, potentially injuring restrained occupants. A 2013 study found that the likelihood of driver fatality increased significantly, 119% to 197%, with the combination of unrestrained rear seat passengers (*8*). The conclusions from this study suggested that future advances to front row passive safety systems (e.g., multi-stage airbag deployment) must be adapted to take into account the effect of unrestrained rear seat passengers, whose usage was only recorded to be 53% (*8*). Research conducted in New Jersey between 2004 and 2006 showed that rear seat occupants' seat belt use was highly significant to injury severity of a belted driver (*9*). A 2005 study estimated that the number of drivers that suffered fatalities or were seriously injured would decrease by approximately 25% if rear seat occupants wear seat belts (*10*). Additionally, the number of front seat passengers that suffered fatalities or were seriously injured was estimated to decrease by 28% if previously unbelted rear seat occupants had worn seat belts (*10*). This demonstrates that the wearing of seat belts by previously unbelted rear seat passengers is effective in reducing not only injuries to the rear seat passengers, but also injuries to front seat occupants (*10*).

The literature on this topic indicates that rear seat belt use is essential for reducing injury severity for both passengers and drivers of vehicles involved in collisions. Public awareness about the importance of seat belts is high in the United States, with 88.2% viewing driving without wearing a seat belt as unacceptable (*11*). However, some occupants in vehicles still choose to not wear any safety restraints—17.8% of survey respondents admitted that they had driven without wearing their seat belt in the past 30 days (*11*). Almost every state has passed legislation requiring the use of seat belts for all adults inside a moving vehicle, with New Hampshire being the only state that does not require seat belt use for adults (*12*). However, only about a third of these states consider it a primary offense to not wear proper safety restraints in the rear seats (*12*). The states that have passed primary laws for rear seat occupants can be

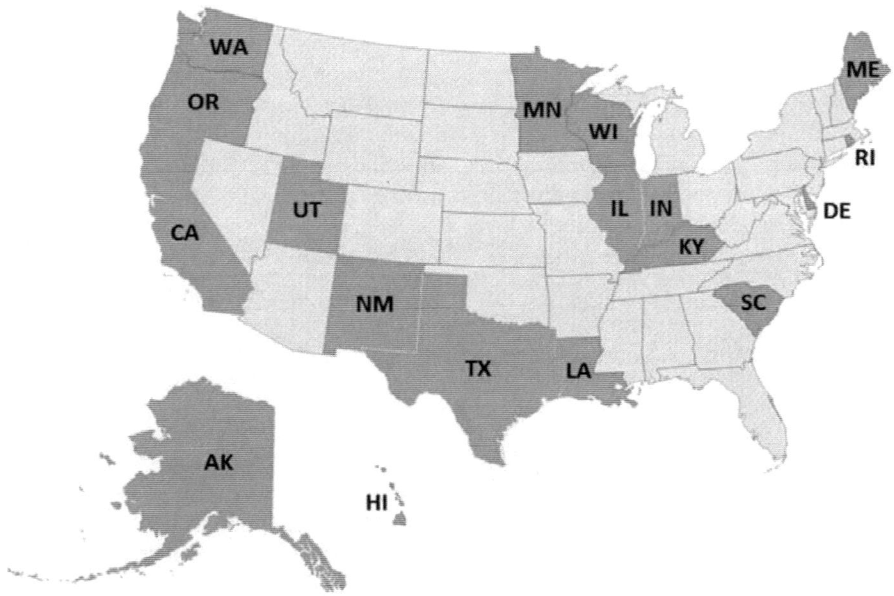

Figure 1. States with primary seat belt laws for all rear seat passengers.

seen in Figure 1. Some other states (Florida, Georgia, Kansas, New Jersey, New York, and West Virginia) only have primary laws passed for those under the age of 18 (Figure 1).

Seat Belt Laws' Effect on Seat Belt Use

The NHTSA conducted surveys in randomly selected roadway sites and reported on observed seat belt usage in all seating positions of motor vehicles. In 2009, the national average for rear seat belt usage was 70%; 14 points lower than front seat occupants (*13*). From 2005 to 2009, states with legislation requiring the use of rear seat belts had, on average, an increase in usage of approximately 14% when compared with states without rear seat belt requirements (*13*). As of November 2016, there are 18 states that have primary restraint laws that require rear seat belt use for all adult passengers (*12*).

The results of a 2016 study that examined seat belt use by teenagers showed that only 51% of high school students studied in 38 states reported always wearing a seat belt when riding as a passenger; prevalence varied from 32% in South Dakota to 65% in Delaware (*5*). Seat belt use was 11% lower in states with secondary enforcement seat belt laws compared with states with primary enforcement laws (*5*). The introduction of seat belt laws may cause survivors of collisions to falsely claim proper belt use when they were not restrained (*14*). However, even when controlling for the potential impact of misreporting seat belt usage, seat belt laws were still shown to be effective at reducing fatalities (*14*). Seat belt laws were also found to have a varying effect on usage among members of different socio-economic classes (*15*). Specifically, primary enforcement of seat belts laws was found to have a stronger effect on those with fewer years of education and those in lower socio-economic classes (*15*). The results of a socio-economic study indicate that differences in seat belt use among various economic classes and education levels would be further mitigated if all states were to start primary enforcement of seat belts (*15*).

Summary of Literature Review

Overall, previous studies found that seat belts are effective for reducing the number of fatalities and injury severity. Additionally, previous studies have found that practicing standard enforcement of seat belt laws has an impact on the usage of proper restraints. Therefore, standard enforcement is likely to result in a reduction of fatalities and injury severity.

Even though there is an abundance of literature on the effects of seat belts laws, seat belt use, and the effect of seat belts on injury severity, there are some limitations for the purposes of this study. Some of the research examined in this literature review was conducted in areas that may not have comparable driver behavior and culture to that of the United States, such as Japan and Norway. Most seat belt research specifically focused on front seat occupants of vehicles. There is a limited amount of completed research that focuses on rear seat belt usage and the potential injury severity reduction as a result of proper rear seat restraint usage. There is also no specific methodology for predicting the amount of fatalities that could have been prevented had there been primary enforcement of rear seat belts in any given state. This paper addresses the need for further research into increased rear seat belt usage and provides a methodology about how to estimate the potential reduction in fatalities as a result of standard enforcement of rear seat belt usage.

Methodology

The FARS database was used to collect data on fatal collisions for all 50 states and the District of Columbia. Variables chosen for data collection from the FARS database included state, year of collision, restraint system use, seating position, and injury severity. Data were collected from the years 2011 to 2015. At the time of data collection, this timeline was the five most recent years available for download. The collected data set was then filtered to only include data for rear seat passengers who were involved in a fatal collision, regardless if they suffered a fatality or not. Involvement in a fatal collision does not mean the rear seat passenger has to suffer a fatality, only that the collision was serious and violent enough to cause a fatality. The fatality could have been a different rear seat occupant or a front seat occupant. These rear seat occupants were considered to be unrestrained if the restraint system use given by FARS was "none used." Other rear seat occupants included in the data set were considered to be restrained if the restraint system use given by FARS was "shoulder and lap belt used," "child restraint system," or "booster." Unknown restraint use data were excluded from the data set. These excluded data comprised fewer than 1% of the total data collected. The filtered data set was then compiled by state to show the statistics of those involved in fatal collisions while in the rear seat.

Next, the restrained and unrestrained fatality rates were calculated for states that did not practice standard enforcement. To calculate the restrained fatality rates, the total number of rear seat fatalities who were restrained was divided by the total number of rear seat occupants who were restrained and involved in a fatal collision. To calculate the unrestrained fatality rates, the total number of rear seat fatalities who were unrestrained was divided by the total number of rear seat occupants who were unrestrained and involved in a fatal collision. This was done for every state without a standard enforcement rear seat belt law. The values for restrained rear seat occupants are shown in Table 1 and the values for unrestrained rear seat occupants are shown in Table 2. Table 2 also shows these states' ratio of unrestrained to restrained fatalities. The values in these two tables were used to calculate predictions for the number of rear seat fatalities which could have been prevented in states without a standard enforcement law (Tables 1 and 2).

The equation used to calculate how many rear seat fatalities could have been prevented with 100% usage of proper restraints is:

$$\text{Rear Seat Fatalities Prevented}_{100\%\text{usage}} = \textit{Unrestrained Fatalities} - (\textit{Total Unrestrained Occupants} \times \textit{Fatality Rate of Restrained Occupants})$$

To predict how many rear seat fatalities could have been prevented with 100% usage of proper restraints, the total number of unrestrained occupants involved in a fatal collision was multiplied by the calculated fatality rate for restrained occupants in a fatal collision. This value would be a prediction for how many fatalities could be expected if all of the unrestrained passengers had instead been restrained. Even though all of these rear seat passengers would now be restrained, the collected data set indicated that between 8% and 20% (this value varied by state) of rear seat passengers would still not survive even if they were properly restrained. The calculated amount of expected fatalities would then be subtracted from the actual amount of fatalities of unrestrained occupants in fatal collisions. The resulting value would be the estimation for the total amount of rear seat fatalities that could have been prevented with 100% usage of seat belts by rear seat occupants in the corresponding state.

The equation used to calculate how many rear seat fatalities could have been prevented by using the ratio of unrestrained to restrained fatalities in states that had a primary rear seat belt law is:

$$\text{Rear Seat Fatalities Prevented}_{\text{Unrestrained:Restrained}} = \textit{Unrestrained Fatalities} - (\textit{Restrained Fatalities} \times \textit{Ratio for Unrestrained to Restrained Fatalities in States with Primary Law})$$

The collected data set was then used to determine an estimate for the number of fatalities that could have been prevented had a primary rear seat belt law been in effect in a state—accounting for a more realistic scenario for rear seat belt usage (less than 100%). The first of two methodologies used the ratio of unrestrained to restrained fatalities in states that had a primary rear seat belt law. Once this ratio was determined, it was multiplied by the total number of restrained rear seat occupants who suffered fatalities in fatal collisions for each state without a primary law. This calculated a prediction for the number of unrestrained fatalities a certain state, without standard

Table 1. Restrained Rear Seat Occupants Involved in Fatal Collisions for States without Standard Seat Belt Enforcement

State	Restrained fatalities	Restrained non-fatalities	Restrained fatality rate (%)	Restrained survival rate (%)
Alabama	86	503	14.6	85.4
Arizona	63	640	9.0	91.0
Arkansas	46	334	12.1	87.9
Colorado	51	375	12.0	88.0
Connecticut	22	131	14.4	85.6
Washington, D.C.	1	11	8.3	91.7
Florida	199	1,812	9.9	90.1
Georgia	132	862	13.3	86.7
Idaho	22	146	13.1	86.9
Iowa	37	99	27.2	72.8
Kansas	43	241	15.1	84.9
Maryland	40	270	12.9	87.1
Massachusetts	17	120	12.4	87.6
Michigan	70	478	12.8	87.2
Mississippi	61	299	16.9	83.1
Missouri	57	506	10.1	89.9
Montana	14	126	10.0	90.0
Nebraska	26	104	20.0	80.0
Nevada	42	220	16.0	84.0
New Hampshire	9	58	13.4	86.6
New Jersey	55	434	11.2	88.8
New York	83	587	12.4	87.6
North Carolina	118	966	10.9	89.1
North Dakota	13	52	20.0	80.0
Ohio	66	666	9.0	91.0
Oklahoma	63	447	12.4	87.6
Pennsylvania	71	587	10.8	89.2
South Dakota	11	65	14.5	85.5
Tennessee	97	620	13.5	86.5
Vermont	6	32	15.8	84.2
Virginia	61	274	18.2	81.8
West Virginia	20	189	9.6	90.4
Wyoming	19	85	18.3	81.7

enforcement, would expect had there been standard enforcement. This predicted amount of unrestrained deaths in fatal collisions would then be subtracted from the actual number of unrestrained deaths to determine an estimate for the number of fatalities that could have been prevented had a primary rear seat belt law been in effect. This methodology directly compares each of the 32 states (and the District of Columbia) that do not have standard seat belt enforcement for rear seat passengers to the 18 states that do practice standard enforcement.

The equation used to calculate how many rear seat fatalities could have been prevented with the NHTSA's observation of a 14% increase in rear seat belt use is:

$$\text{Rear Seat Fatalities Prevented}_{14\%\text{usage increase}} = (\textit{Total Unrestrained Occupants} \times 14\%) \times \textit{Restrained Survival Rate}$$

The second methodology to determine an estimate for the number of fatalities that could have been prevented had a primary rear seat belt law been in effect incorporated NHTSA's observation of a 14% increase in rear seat belt use for states enforcing a primary law. The total number of a state's unrestrained rear seat passengers in a fatal collision was multiplied by 14% to determine the expected number of people that would instead have been restrained because of the effect of standard enforcement. This determination for the number of people that would have been restrained was then multiplied by each state's calculated survival rate of restrained rear seat occupants involved in fatal collisions. The resulting value would be an estimation of fatalities prevented because of a 14% increase in rear seat belt use.

Overall, this study is limited because the data are based on reported fatal collisions. Additionally, the data are reliant on accurate reports for seat belt usage. However, both of these limitations are expected to have a minimal, if not negligible, effect on the final results.

Results

100% Usage of Seat Belts

From 2011 to 2015, there was a total of 8,444 unrestrained rear seat passengers that were involved in a fatal

Table 2. Unrestrained Rear Seat Occupants Involved in Fatal Collisions for States without Standard Seat Belt Enforcement

State	Unrestrained fatalities	Unrestrained non-fatalities	Unrestrained fatality rate (%)	Unrestrained survival rate (%)	Unrestrained to restrained fatalities ratio
Alabama	155	278	35.8	64.2	1.80
Arizona	145	298	32.7	67.3	2.30
Arkansas	66	125	34.6	65.4	1.43
Colorado	97	158	38.0	62.0	1.90
Connecticut	25	48	34.2	65.8	1.14
Washington, D.C.	1	3	25.0	75.0	1.00
Florida	343	631	35.2	64.8	1.72
Georgia	168	306	35.4	64.6	1.27
Idaho	35	73	32.4	67.6	1.59
Iowa	54	81	40.0	60.0%	1.46
Kansas	66	114	36.7	63.3	1.53
Maryland	66	93	41.5	58.5	1.65
Massachusetts	44	57	43.6	56.4	2.59
Michigan	122	181	40.3	59.7	1.74
Mississippi	135	265	33.8	66.3	2.21
Missouri	125	228	35.4	64.6	2.19
Montana	47	98	32.4	67.6	3.36
Nebraska	48	82	36.9	63.1	1.85
Nevada	39	70	35.8	64.2	0.93
New Hampshire	9	21	30.0	70.0	1.00
New Jersey	98	158	38.3	61.7	1.78
New York	148	295	33.4	66.6	1.78
North Carolina	209	344	37.8	62.2	1.77
North Dakota	29	44	39.7	60.3	2.23
Ohio	147	281	34.3	65.7	2.23
Oklahoma	125	210	37.3	62.7	1.98
Pennsylvania	150	238	38.7	61.3	2.11
South Dakota	25	49	33.8	66.2	2.27
Tennessee	137	216	38.8	61.2	1.41
Vermont	7	12	36.8	63.2	1.17
Virginia	135	192	41.3	58.7	2.21
West Virginia	28	65	30.1	69.9	1.40
Wyoming	33	69	32.4	67.6	1.74

collision in a state without a standard enforcement seat belt law. Of these passengers, 3,061 suffered fatalities as a result of these collisions, yielding a fatality rate of 36.3%. If all 8,444 of these rear seat passengers had been using proper restraints, only 1,071 of these passengers would be expected to suffer fatalities as a result of these collisions. A total of 1,071 fatalities for all 8,444 passengers is a fatality rate of just 12.7%. This is a fatality rate reduction of 23.6% and a total difference of 1,990 people that may have survived had there been 100% usage of seat belts.

Ratio of Unrestrained to Restrained Fatalities

The ratio of unrestrained to restrained rear seat fatalities was 1.78 in states without standard seat belt enforcement. States that practice standard rear seat belt enforcement had an unrestrained to restrained rear seat fatalities ratio of 1.33. This indicates that, for states without a standard enforcement law, there are on average 0.45 more unrestrained fatalities for every properly restrained rear seat fatality. The ratio of 1.33 unrestrained to restrained rear seat fatalities was applied to the 2011 to 2015 total restrained deaths for each state without a standard enforcement seat belt law. Overall, it is estimated that 772 fatalities may have been prevented in these states without standard enforcement from 2011 to 2015. Excluding the negative values (as it would be unreasonable to conclude that a standard enforcement seat belt law would increase fatalities), this estimation rises to 805 fatalities prevented from 2011 to 2015 (Table 3).

14% Increase in Seat Belt Use

The NHTSA's observation of a 14% increase in rear seat belt usage was applied to the selected states' total number of unrestrained rear seat passengers that were involved in fatal collisions. This methodology essentially shifts 14% of the unrestrained passengers to the restrained category. The state's corresponding fatality rate for restrained rear

Table 3. Ratios of Unrestrained to Restrained Fatalities for States with and without Primary Rear Seat Belt Laws in Effect

Primary rear seat belt enforcement	Total restrained fatalities (2011–2015)	Total unrestrained fatalities (2011–2015)	Ratio of unrestrained to restrained fatalities
States without primary enforcement	1,721	3,061	1.78
States with primary enforcement	1,490	1,987	1.33

seat passengers is then applied to these newly added restrained passengers. The resulting value from this methodology is the estimate for the number of fatalities that would have been prevented had there been a 14% increase in rear seat belt usage. This methodology predicted that 1,032 fatalities, or 206 fatalities per year, would have been prevented as a result of the increase in seat belt use (Table 4).

Comparing Rear Seat Fatality's Age to the Corresponding Driver's Age

In addition to the calculations performed to estimate the fatality reduction with standard enforcement, the research team compared the ages of the unrestrained rear seat fatalities to their corresponding driver's age. Unrestrained rear seat fatalities' ages were grouped into five age ranges:

- 0 to 7 years old (very young children, still needing additional child restraints)
- 8 to 15 years old (young children, no longer needing additional child restraints)
- 16 to 17 years old (teenagers and young drivers, most likely still in high school)
- 18 to 25 years old (young adults, college students, or recent high school graduates)
- 25 years + (adults, parents)

Corresponding drivers' ages were grouped into similar age ranges (16 to 17 years old, 18 to 25 years old, and 25 years +). The total amount of rear seat fatalities was then found for each age range comparison. The percent of unrestrained rear seat fatalities by driver age group was then calculated for each age range (for example, 71% of all 0- to 7-year-old unrestrained rear seat fatalities in the U.S. from 2011 to 2015 were in a vehicle being driven by someone 25 years or older) (Figure 2).

Conclusions and Implications

This study examined the relationship between seat belt use and fatality rates for occupants in rear seats using various methods to determine how many fatalities may have been prevented if a state enacted a standard enforcement law for seat belt use. Three predictions for fatality prevention were calculated for each state that did not have a standard enforcement law in the years 2011 to 2015. The predictions included a best-case scenario, assuming 100% seat belt use, and more realistic scenarios, including a 14% increase in seat belt use and the change of unrestrained to restrained fatality ratios to 1.33. The more realistic scenarios were calculated because it is unreasonable to expect 100% seat belt usage by rear seat passengers, based on the experience of other states who have exacted similar laws.

The best-case scenario resulted in an estimated fatality prevention of 1,990 rear seat passengers from the years 2011 to 2015. The more realistic scenarios yielded a smaller, but still substantial, average of 902 occupants saved in the same time period (1,032 occupants with the increase in usage and 772 occupants with the ratio change).

To increase the reliability and confidence of the realistic results and their implications, this study incorporated multiple methodologies to estimate annual fatality preventions. The methodology that assumed a 14% increase in rear seat belt use yielded less conflicting results and a higher average for the estimation of fatality prevention. The other realistic methodology that compared the state's ratio of unrestrained to restrained fatalities yielded a lower average for the estimation of fatality prevention due to standard enforcement. The slightly lower average value can be explained by the states that show negative values for fatalities prevented. These states (Connecticut, Georgia, Nevada, New Hampshire, and Vermont) all had unrestrained to restrained fatality ratios lower than the average ratio for states that practice standard enforcement. The low ratio in these states could be a result of a small sample size or because existing driving behavior and culture result in an above average seat belt usage rate. Although this individual method predicts an increase in fatalities, the actual changes are expected to be improvements.

The results of the comparison between the age of a rear seat fatality and their corresponding driver's age indicate that most fatalities of young children occur with a driver that is more than 25 years old. Additionally, the percentage of fatalities for teenagers and young adults are much higher in the same drivers' age range. For

Table 4. Estimations for the Number of Fatalities Prevented from 2011 to 2015

State	Estimated number of fatalities prevented (2011–2015)		
	100% seat belt usage	If the state's unrestrained to restrained ratio was 1.33*	14% increase in rear seat belt usage
Alabama	92	41	52
Arizona	105	61	56
Arkansas	43	5	24
Colorado	66	29	31
Connecticut	15	–4	9
Washington, D.C.	1	0	1
Florida	247	78	123
Georgia	105	–8	58
Idaho	21	6	13
Iowa	17	5	14
Kansas	39	9	21
Maryland	45	13	19
Massachusetts	31	21	12
Michigan	83	29	37
Mississippi	67	54	47
Missouri	89	49	44
Montana	33	28	18
Nebraska	22	13	15
Nevada	22	–17	13
New Hampshire	5	–3	4
New Jersey	69	25	32
New York	93	38	54
North Carolina	149	52	69
North Dakota	14	12	8
Ohio	108	59	55
Oklahoma	84	41	41
Pennsylvania	108	56	48
South Dakota	14	10	9
Tennessee	89	8	43
Vermont	4	–1	2
Virginia	75	54	37
West Virginia	19	1	12
Wyoming	14	8	12
5-year total	1,990	772	1,032
Average annual total	398	154	206

*Although this method generates some negative values, the implication in these states would not be an increase in fatalities values are totals and were submitted as shaded cells.

example, 67% of all 18- to 25-year-old rear seat fatalities occur in a vehicle driven by someone 18 to 25 years old. Similarly, for all 16- to 17-year-old unrestrained rear seat fatalities, 34% were in a vehicle driven by another 16- to 17-year-old. The increased percentage of unrestrained fatalities within each young adults' age group indicates that, should a transportation, public health, law enforcement, education, or other interested agency or organization desire to lower unrestrained rear seat fatalities for young adults, the agency should target similarly aged drivers to encourage them to make sure their rear seat passengers are wearing proper restraints. Additionally, should an agency desire to lower unrestrained rear seat fatalities for young children, the transportation agency should target adults and parents to encourage them to make sure their young children in the rear seat are wearing proper restraints for their age and weight.

To further examine the relationship between a state's safe driving culture and their ratio of unrestrained to restrained fatalities, the research team compared each state's unrestrained to restrained fatality ratio to their average seat belt use percentage for all occupants from 2011 to 2015 (*16*). Figure 3 shows the resulting comparison between the state's seat belt usage and their unrestrained to restrained fatalities ratio. A trend can be seen that demonstrates the logical reasoning that an increased seat belt usage rate will tend to have a lower corresponding unrestrained to restrained fatality ratio. This helps explain why states, such as Georgia and Nevada, have such low ratios despite not having standard enforcement

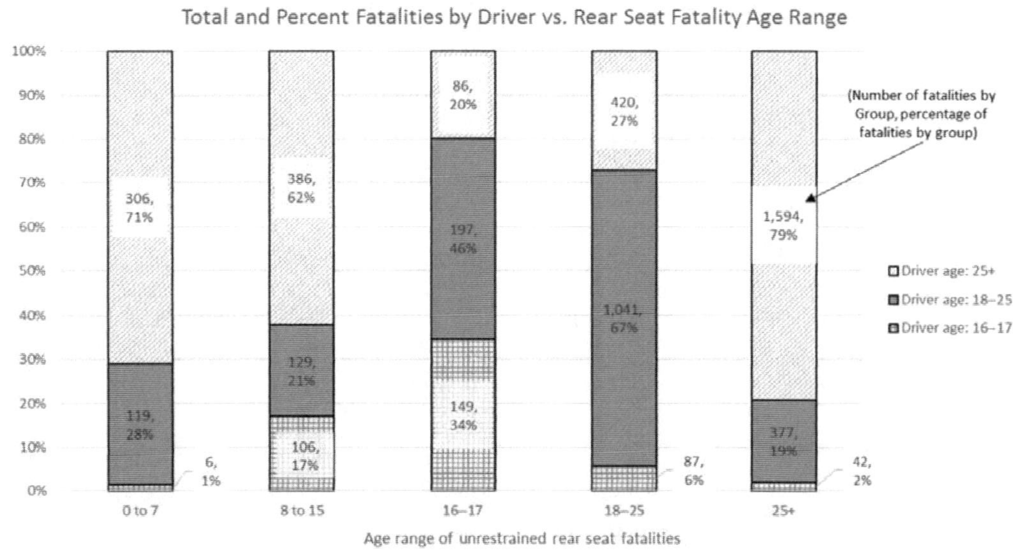

Figure 2. Graph comparing the age of unrestrained rear seat fatalities to their corresponding driver's age.

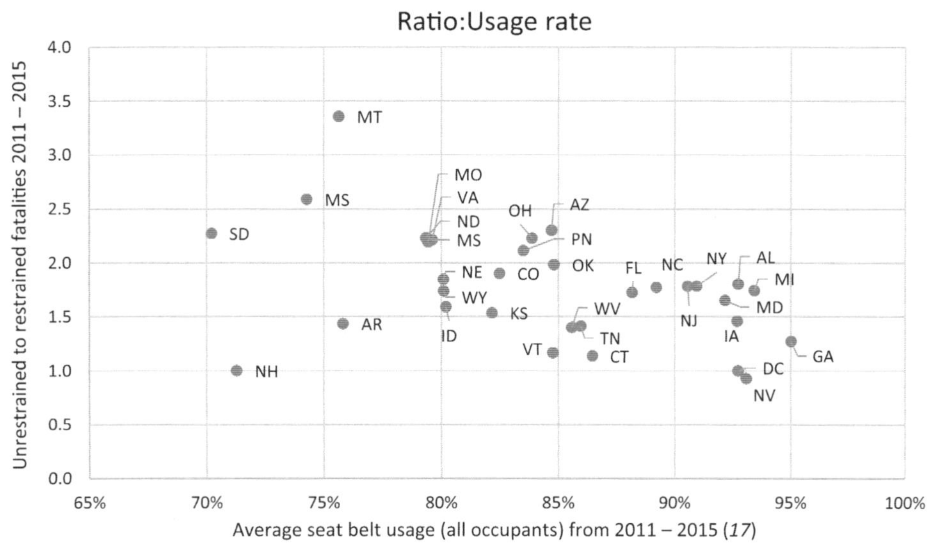

Figure 3. Graph of average seat belt use for front seat occupants compared with the state's corresponding ratio of unrestrained to restrained fatalities. Seat belt usage percentages are representative of daytime conditions.

of rear seat belts. Further study could be done to examine what these states are doing to maintain such a high percentage of seat belt use despite not having a primary enforcement law for rear seats (Figure 3).

The economic impacts of reducing injury severity and fatality prevention can be estimated using the difference between cost per crash of a fatal collision and a type-A injury collision. This value will vary state by state, but as an example, the economic impact of a standard enforcement law was calculated for the state of North Carolina. The North Carolina value for the difference between a fatal collision and a type-A injury collision in crash costs was last reported in 2013 by the North Carolina Department of Transportation to be $9,569,000 (*17*). A prevented fatality because of standard seat belt enforcement can be represented as reducing a fatal collision to a type-A collision, though the injury may be even less severe. Multiplying the crash cost savings due to the reduction from a fatal collision to a type-A collision by the average of the realistic results (12 fatality preventions per year) in North Carolina would result in a yearly saving of $115 million in crash costs. This value does not include any potential savings in costs by reducing non-fatal injury severity in collisions, meaning that the $115 million can be considered a conservative estimation for yearly crash costs savings. This process can be easily repeated for any state that has reported crash costs to help identify an individual state's potential economic impact of a standard enforcement seat belt law. The North Carolina value for the difference in cost between a fatal and type-A collision was applied to the average, realistic 5-year fatality prevention total for every state, 902 fatalities from 2011 to 2015. If the North Carolina fatal crash cost savings are representative of the other states, the national savings are estimated to be $8.6 billion, or $1.7 billion annually.

Recommendations

If a primary law goes into effect in one of the states without standard enforcement, a study could be done to observe the actual impact of the standard enforcement law on seat belt use and any potential drop in fatalities or the ratio of unrestrained to restrained fatalities. The results from such studies can help inform traffic safety decision makers and influence rear seat belt use through public awareness campaigns or law enforcement activities.

Three methods were presented that are incumbent on the prevailing vehicle fleet, highway design standards, vehicle speeds, driver characteristics, and other influencing factors. In states without primary rear seat belt enforcement, 3,061 unrestrained individuals suffered a fatal injury in motor vehicle collisions in the United States from 2011 to 2015. In the most optimistic scenarios (while still accounting for fatalities that occur even while properly restrained) presented, it was estimated that 1,990 lives could have been saved over these 5 years. Under more realistic scenarios, it was estimated that 772 or 1,032 lives could have been saved. Therefore, many states should consider the implementation of a primary enforcement law for rear seat passengers.

Author Contributions

The authors confirm contribution to the paper as follows: study conception and design: Timothy Nye, Daniel Findley; data collection: Timothy Nye, Morgan Sanchez; analysis and interpretation of results: Timothy Nye, Daniel Findley; draft manuscript preparation: Timothy Nye. All authors reviewed the results and approved the final version of the manuscript.

References

1. *Click It or Ticket*. North Carolina Department of Transportation Governor's Highway Safety Program. https://www.ncdot.gov/programs/GHSP/initiatives/clickitorticket/. Accessed July 27, 2017.
2. NHTSA. *Strengthening Seat Belt Use Laws*. DOT HS 810 890W. National Highway Traffic Safety Administration, Washington, D.C., 2008. https://www.nhtsa.gov/DOT/NHTSA/Communication%20&%20Consumer%20Information/Articles/Associated%20Files/810890.pdf.
3. Luchini, D., M. Sammicheli, and C. Cortucci. The Effect of Rear-Seat Overloading in a Car Crash. *The American Journal of Forensic Medicine and Pathology*, Vol. 34, No. 3, 2013, pp. 177–180.
4. McGehee, D., M. Reyes, and D. Marshall. *A Comparative Policy Analysis of Seat Belt Laws*. IADOT Rep No. RB37-013. Iowa Department of Transportation and the Federal Highway Administration, 2014.
5. Shults, R. A., T. M. Haegerich, G. Bhat, and X. Zhang. Teens and Seat Belt Use: What Makes them Click? *Journal of Safety Research*, Vol. 57, 2016, pp. 19–25.
6. Kahane, C. J. Fatality and Injury Reducing Effectiveness of Lap Belts for Back Seat Occupants. *SAE Transactions*, Vol. 96, No. 2, 1987, pp. 416–423.
7. Høye, A. How Would Increasing Seat Belt Use Affect the Number of Killed or Seriously Injured Light Vehicle Occupants? *Accident Analysis & Prevention*, Vol. 88, 2016, pp. 175–186.
8. Bose, D., C. Arregui-Dalmases, D. Sanchez-Molina, J. Velazquez-Ameijide, and J. Crandall. Increased Risk of Driver Fatality Due to Unrestrained Rear-Seat Passengers in Severe Frontal Crashes. *Accident Analysis & Prevention*, Vol. 53, 2013, pp. 100–104.
9. Nukenine, S., and J. Daniel. Impact of Belt Use by Rear-Seat Occupants on Injury Severity of Belted Drivers in New Jersey. *Transportation Research Record: Journal of the Transportation Research Board*, 2011. 2265: 91–99.
10. Shimamura, M., M. Yamazaki, and G. Fujita. Method to Evaluate the Effect of Safety Belt Use by Rear Seat Passengers on the Injury Severity of Front Seat Occupants. *Accident Analysis & Prevention*, Vol. 37, No. 1, 2005, pp. 5–17.
11. AAA foundation for Traffic Safety. *2015 Traffic Safety Culture Index*. https://www.aaafoundation.org/sites/default/files/2015_TSCI.pdf. Accessed July 27, 2017.
12. Centers for Disease Control and Prevention. *Primary Enforcement of Seat Belt Laws*. https://www.cdc.gov/motorvehiclesafety/calculator/factsheet/seatbelt.html. Accessed July 27, 2017.
13. Pickrell, T. M., and T. J. Ye. Occupant Restraint Use in 2009: Results From the National Occupant Protection Use Survey Controlled Intersection Study. *Annals of Emergency Medicine*, Vol. 57, No. 6, 2011, p. 688.

14. Cummings, P., J. D. Wells, and F. P. Rivara. Estimating Seat Belt Effectiveness Using Matched-Pair Cohort Methods. *Accident Analysis & Prevention*, Vol. 35, No. 1, 2003, pp. 143–149.
15. Harper, S., E. C. Strumpf, S. Burris, G. D. Smith, and J. Lynch. The Effect of Mandatory Seat Belt Laws on Seat Belt Use by Socioeconomic Position. *Journal of Policy Analysis and Management*, Vol. 33, No. 1, 2013, pp. 141–161.
16. *Seat Belt Use in 2015 – Use Rates in the States and Territories*. DOT HS 812 274. National Highway Traffic Safety Administration, Washington, D.C., 2016.
17. Robinson, B. *2013 Standardized Crash Cost Estimates for North Carolina*. North Carolina Department of Transportation, 2014.

The Standing Committee on Occupant Protection (ANB45) peer-reviewed this paper (18-01051).

TRR
JOURNAL OF THE TRANSPORTATION RESEARCH BOARD

Article

Transportation Research Record
2018, Vol. 2672(33) 78–88

DOI: 10.1177/0361198118786245
journals.sagepub.com/home/trr
SAGE

Distracted Driving Behaviors and Beliefs among Older Adults

Linda Hill[1], Sara Baird[1], Jessa K. Engelberg[1,2], Jacob Larocca[1], Uns Alwahab[1], Jasmine Chukwueke[1], Anne-Marie Engler[1], Jana Jahns[1], and Jill Rybar[1]

Abstract

Distracted driving, and especially cell phone use, is a prevalent contributor to crashes. Older drivers have an increased risk of committing safety errors while driving, especially with distraction. The objective of this study was to characterize phone-related distractions in older drivers (age > 65) and identify intervention strategies likely to reduce cell distraction. A 64-question survey was offered online and on paper. A distracted driving scale (DDS) was created by summing responses on 11 distracted driving questions related to phone use (possible score range 0 to 44). Linear regression was performed to identify variables associated with a higher DDS score. A total of 363 older drivers completed the survey; the mean age was 73 and 56% were female. 60% of older adults reported using their cell phone while driving at least some of the time. Participants perceived their own ability as capable or very capable when driving and using: handheld phone (40%); hands-free phone (78%); other tasks (38%) while driving. 32% of older adults who drive minors reported driving while distracted. 30% of those who work felt obligated to take work-related calls. Variables associated with distracted driving include younger age, driving more miles, perceived hands-free skill, smart phone ownership, and being employed or self-employed. State laws and potential loss of insurance coverage with distracted driving were cited as effective penalties. Older adults are engaging in distracted driving, including with minors in the car. Education, state laws, and altering insurance coverage may mitigate this behavior.

In the U.S. in 2015, 3,477 people were killed and about 391,000 were injured due to distracted driving (DD); this accounts for 6.7% of all fatal crashes that year (*1*). DD can take many forms, including talking or texting on a cell phone, reading, using a navigation device, eating or drinking, grooming, interacting with children, and talking to passengers. Research overwhelmingly supports a detrimental relationship between DD and crash risk (*2*).

Of all the forms of DD, cell phone use is especially prevalent in the U.S. and is involved in a sizeable portion of motor vehicle crashes. As of January 2017, 95% of Americans report owning some kind of mobile phone and 77% report owning a smart phone (*3*). In a 2011 study, 69% of users report talking on the phone and 31% report texting or checking email while driving (*4*). This translates to an estimated 6.9% of vehicles whose drivers were using some type of phone (handheld or hands-free) at any typical daylight moment in 2015 (*5*).

Studies find people are four to six times more likely to be involved in a car crash while using a cell phone (*6*, *7*). Though many drivers transition to hands-free in states with restrictive DD laws (*8*), the likelihood of being involved in a motor vehicle collision is not reduced when using a hands-free device (*9*, *10*). The delayed reaction time involved with cell phone usage (both hands-free and handheld) is the equivalent of driving with a blood alcohol content of 0.04–0.12. (*11*, *12*). Other distractions that take eyes, hands and mind off driving have also been shown to increase crash risk, including the use of navigation devices (*13*, *14*), even if they are voice-controlled (*15*).

Despite evidence of the risk of driving distracted, individuals tend to overestimate their driving capabilities. When drivers in one study were asked if talking on the phone or sending messages made a difference in their driving ability, 54% of participants said that talking on a handheld device made no difference, and 25% said texting or sending messages made no difference to their driving performance (*16*). In another study, middle-aged participants rated their abilities to talk while driving as competent or very competent when talking on a handheld phone (52%), hands-free phone (89%) or texting (8%) (*17*).

[1]Department of Family Medicine and Public Health, University of California, San Diego, CA
[2]West Health Institute, La Jolla, CA

Corresponding Author:
Address correspondence to Linda Hill: lhillbaird@gmail.com

Approximately 75% of adults aged 75 and older are active drivers (*18*). While research to date has largely focused on distraction among younger drivers, there is less known about the driving habits of older adults, especially regarding cell phone use. Even without distraction, older drivers are at an increased risk of committing safety errors while driving (*19*) and performed worse than middle-aged drivers when distracted (*20*, *21*).

The effect of distraction mitigation strategies is unclear in this population. In studies of all drivers, restrictive driving laws were associated with decreased overall phone use, while rates of texting remained unchanged (*22*). Other studies show that distraction mitigation strategies, such as the cost of citations, points on the driver's license and insurance-related penalties, may be effective in middle-aged and older populations (*17*, *23*, *24*). More data are needed to understand how older adults distracted drive, how they perceive their behavior, and what methods would most effectively reduce the dangers to older adult drivers.

Two previously published DD surveys conducted by this group's Training, Research and Education for Driving Safety (TREDS) program evaluated driver behavior and attitudes regarding cell phone use among college-aged (*25*) and middle-aged (*17*) adults. In an extension of that work, the objectives of this exploratory study were to examine (*a*) incidence of various phone-related distractions experienced by older adults (age $\geq$ 65) while driving; (*b*) older adults' beliefs about DD; (*c*) predictors of DD among older adults; and (*d*) potential of specific intervention strategies among older adults to reduce cell phone-related distraction while driving.

Methods

This study and all materials were approved by the Institutional Review Board of the University of California, San Diego.

Survey Development

A DD survey was previously developed and validated in college (*25*) and middle-aged (*17*) populations. The DD survey was modified to target older adults based on expert feedback considering the driving patterns of older adults. The survey focused on five categories of questions: (*a*) demographics, including personal health issues; (*b*) personal cell phone behaviors and observed behaviors of others; (*c*) perceived effectiveness of potential interventions; (*d*) participants' perceived driving ability and ability to multitask; and (*e*) cell phone behavior when driving with minors.

The study was initially piloted with 20 San Diego older adults using a mix of random-sampling and convenience sampling, incentivized with a $20 gift card. The pilot feedback was used to assess face validity and modify the survey accordingly. After edits, additional feedback was obtained with five additional volunteers and the survey was finalized.

Survey Administration

Eligibility criteria required that participants were (*a*) 65 years of age or older, (*b*) using a cell phone at least once a week, (*c*) driving a vehicle at least once a week, (*d*) a U.S. resident, and *e*) English speaking.

The final survey was implemented between May 1 and November 30, 2016. Up to 64 questions were answered, based on skip-logic embedded in the survey. Two reliability questions were included. On average, participants completed the survey in 15 to 20 minutes.

Surveys were completed either online or on paper. Recruitment was done through flyers in participating physician offices, social media, internet posting on the TREDS website, regional newspaper advertisements, and various online and email listservs. Paper surveys were distributed in physician offices in San Diego; participants were given a paper copy of the survey and an addressed stamped envelope to return the survey to research staff. Online participants accessed the survey through Surveymonkey.com. Both surveys included initial eligibility questions, and all participants were invited to participate in an incentive lottery. If interested, participants were directed to complete a second separate survey to provide necessary contact information; no identifying information was linked to survey results.

Measures

Demographics included age, gender, income, ethnicity, employment status, living location, marital status, and cell phone ownership. Respondents were also asked to report common health conditions, with a focus on conditions that may impact driving performance. Finally, participants were asked to rate their quality of day and night vision.

Survey questions also assessed the frequency of selected driving behaviors and attitudes, including behaviors around DD; when feasible these questions were measured with Likert-scale response options. Based on existing research as well as the expertise of the authors, questions were designed to explore multiple facets of DD.

From the survey, a 12-item distracted driving scale (DDS) was created by summing responses on DD questions in terms of frequency of behavior (Table 1). For example, "How much time do you spend talking on a handheld cell phone?" had five Likert-scale response options (0 = never, 1 = rarely, 2 = sometimes, 3 = often, 4 = frequently). The DDS was based on items used in the survey of middle-aged adults (*17*).

Data Analysis

Analyses were run in SPSS v. 22.0 and R v. 3.3.2 using a *p*-value of .05 to determine significance. An exploratory factor analysis was conducted to determine if the scale had multiple factors (e.g., hypothesized texting behaviors vs. talking behaviors). One DDS question was found to have a borderline

Table 1. 12-item DDS[a]

DD1: Of the time that you spend driving on an average day, how much time do you spend talking on a HANDHELD cell phone?
DD2: Of the time that you spend driving on an average day, how much time do you spend using a HANDS-FREE device with your cell phone?
DD3: Of the time that you spend driving on an average day, how much time do you spend TEXTING with your cell phone?
DD4: How often do you talk on a cell phone (handheld or hands-free) while driving with passengers age 11 years old or younger?[b]
DD5: How often do you TEXT while driving with passengers age 11 years old or younger?[b]
DD6: How often do you talk on a cell phone (HANDHELD or HANDS-FREE) while driving with passengers between the ages of 12–17 years?[b]
DD7: How often do you TEXT while driving with passengers between the ages of 12–17 years?[b]
DD8–12: While driving, how frequently do you use these devices while the vehicle is moving?
DD8: Navigation device (e.g., TomTom, Garmin, smartphone map application)
DD9: Portable music player (e.g., iPod)
DD10: Laptop/Tablet
DD11: Smartphone applications (e.g., games, email, music)
DD12: Other electronic devices[c]

[a]Answer options: 0= never (only emergency/911 calls); 1= rarely (less than 10% of the time); 2= sometimes (about 25% of the time); 3= often (about 50% of the time); 4= frequently (greater than 75% of the time).
[b]Only includes participants who reported driving with minor passengers.
[c]This question was removed from final analysis.

loading (0.349) and communality (0.12); when removed, the eigenvalue dropped from 6.60 to 6.50 but the amount of variance explained increased from 55% to 59%. Based on these results, the last DDS question ("other electronic devices") was excluded from further analysis.

The final 11-point DDS had a possible score range of 0 to 44. Initial analysis included descriptions of participant characteristics, frequency of engaging in different DD behaviors, and frequency of attitudes and beliefs pertaining to DD. Chi-squared tests and one-sided Wilcox Rank-Sum tests were performed to determine the association between DD scores of $\geq$1 and various self-reported health conditions.

A linear regression model was run in R where the DDS was the outcome; the full model included demographic variables and expected correlates of DD behaviors. The following measures were considered in the regression analysis: age, ethnicity, income, smartphone ownership, employment status, days driving per week, time spent driving per day, perceived hands-free skill, perceived ability to multitask while driving, perceived ability to drive safely compared to other drivers, and perceived skill at driving safely while using a handheld phone or texting.

While each individual question in the DDS is ordinal, the scale was treated as continuous. Because the DDS scores were right-skewed, the data were transformed using square root of sum to improve the quality of the analysis to better approximate a normal distribution for analysis while providing interpretable estimates. The appropriateness of this method was confirmed by running a bivariate analysis and plotting the response variables in a histogram. Variables of interest, as outlined above, were identified and tested via a step function in R to identify significant predictors for the regression model, and variables that did not significantly contribute to the DDS model were excluded. Following analysis, diagnostic plots were created to ensure the validity of the linear regression. Logistic regression was also performed with similar results, using DDS = 0 versus >1.

Results

A total of 397 participants accessed the survey and 363 (91.4%) met eligibility requirements. Of those who were ineligible, six (17%) were under 65, twenty-six (77%) used a cell phone less than once per week, and two (6%) were not U.S. residents. More than 90% of surveys were completed online.

Participant Characteristics

Descriptive participant information is summarized in Table 2.

Respondents were 56% female, 87% white non-Hispanic, and the mean age was 73, with a range of 65 to 94 years. Almost 24% reported an income level under $50,000 per year, and 25% reported being currently employed or self-employed. 83% were smartphone owners. Of note, not all participants chose to answer demographic information; approximately 90% of respondents answered each demographic question with the exception of medical conditions, which was answered by only 62% of respondents. Over half of those answering reported at least one medical condition (e.g., arthritis (50%), high blood pressure (47%), diabetes (16%)), which could potentially affect one's driving ability. Several participants listed medical conditions that impair vision, including glaucoma (4%) and/or macular degeneration (4%). About 2% reported a diagnosis of dementia, Parkinson's or stroke. While all respondents reported at least fair day vision for driving, 7% reported poor or somewhat poor night vision for driving.

Table 2. Participant Descriptives

	N(%)/mean(SD)
Gender (*n* = 339)	
Male	147 (43.4%)
Female	192 (56.6)
Ethnicity (*n* = 322)	
White	283 (87.9%)
Hispanic	18 (5.6)
Other	21 (6.5)
Income (*n* = 267)	
Less than $50,000	63 (23.6%)
$50,000–$99,999	93 (34.8)
$100,000–$149,999	66 (24.7)
$150,000 or more	45 (16.9)
Smartphone owner (*n* = 344)	285 (82.8%)
Employment status (*n* = 348)	
Employed	64 (18.4%)
Not employed	259 (74.4)
Self employed	25 (7.2)
Living location (*n* = 335)	
Single family home	263 (78.5%)
Condominium/apartment	40 (11.9)
Senior residential community	27 (8.1)
Other	5 (1.5)
Marital status (*n* = 337)	
Single	26 (7.7%)
Married/domestic partnership	201 (59.6)
Widowed	62 (18.4)
Divorced	48 (14.2)
Age	
All (*n* = 331, range 65–94 years)	73.1 (6.6)
Female (*n* = 189, 65–89)	72.4 (6.2)
Male (*n* = 142, 65–94)	74.1 (7.1)
Medical conditions (*n* = 226)	
Arthritis	113 (50.0%)
High blood pressure	108 (47.8)
Diabetes	37 (16.4)
Cancer (in the last 2 years)	24 (10.6)
Heart disease	19 (8.4)
Depression	16 (7.1)
Glaucoma	15 (6.6)
Macular degeneration	14 (6.2)
Parkinson's disease	3 (1.3)
Stroke	3 (1.3)
Dementia	2 (0.9)
Epilepsy	1 (0.4)
Quality of day vision (*n* = 334)[a]	4.58 (0.6)
Quality of night vision (*n* = 331)[a]	3.70 (0.9)

[a]Range 1 (poor) – 5 (excellent).

Behavior and Attitudes

The driving behaviors and attitudes of participants are summarized in Table 3.

Participants were generally active drivers, driving an average of nearly 6 days per week and 3 round trips per day. 44% of participants drove more than 50 miles per week. 82% of participants owned a smartphone. All but one participant used their smartphone for talking, though many used their smartphone for other capabilities (i.e., texting [78%], email [64%], browsing internet [56%], navigation [54%], and music [24%]).

Nearly 60% of respondents reported using their phone while driving, including 42% who talk (hands-free or handheld) and 9% who text while driving. 61% of participants use their phone for navigation while driving at least rarely, and 11% of participants use it often or frequently.

When asked about incoming calls, 42% of participants reported that they do not answer calls while driving. The remaining 58% reported accepting calls in at least some capacity while driving: almost 10% tell their caller that they will call right back, 12% reported that they pull over after answering, 12% answer only when they recognize the caller, and 23% answer all calls. Of the participants who take all calls, 35% specifically mentioned a hands-free function in the comment section. Of those who were employed, nearly 30% felt obligated to take work-related calls while driving.

When asked about outgoing calls, 39% of participants make outgoing calls while driving. Of those, 45% are using voice commands on their phone or in the car. The remainder reported using their hands to scroll through their contacts list, use speed dial, or manually enter numbers when they are placing calls.

In addition to using their phone while driving alone or with other adults, participants also engage in these behaviors while driving with minor passengers. A total of 26% of participants reported driving with a minor in the car. Of those who drive with minors, 33% of participants reported using their phone either hands-free or handheld while driving with younger children (under age 11) and about 42% when driving with older children (aged 12 to 17). Though less common, about 1% reported texting when driving minors (regardless of age). 87% reported the minors were their grandchildren.

Participants rated their own driving abilities highly; almost 80% reported that they drive better than their peers. Participants also perceived their own ability as capable or very capable when driving and using a handheld phone (40%), a hands-free phone (78%), or doing other tasks while driving (38%).

Citations and Crashes

The survey included questions about adverse driving events, including citations and crashes. About 18% of participants were involved in collisions in the last 2 years, and 43% were given a non-parking citation in the last 2 years. Though only 10 respondents had received a ticket for talking on the phone

Table 3. Knowledge and Behaviors

Driving knowledge and behavior	N(%)/mean(SD)
Number of years driving (*n* = 326)	56.20 (7.7)
Number of days a week driving (*n* = 344)	5.78 (1.4)
Time spent driving per day (*n* = 345)	0.81 (0.5)
What do you do with your smartphone (*n* = 348)	
Talk	347 (99.7%)
Text	274 (78.7)
GPS/navigation	191 (54.9)
Bluetooth	150 (43.1)
Phone "apps"	186 (53.5)
Email	224 (64.4)
Music	84 (24.1)
Podcasts	49 (14.1)
Browse internet	198 (56.9)
Primary purpose of phone use while driving (*n* = 344)	
Personal	150 (43.5%)
Work-related	19 (5.5)
Critical information (directions, etc.)	24 (7.0)
I never use the phone while driving	143 (40.6)
Other	9 (2.6)
What do you do when the phone rings (*n* = 344)	
Answer; pull over	42 (12.2%)
Answer; tell the caller you will call back	34 (9.9)
Answer; take the call	79 (23.0)
Answer only if recognized number	41 (11.9)
Don't answer	147 (42.7)
Other ("it depends")	1 (0.3)
How do you make calls while driving (*n* = 358)	
Manually enter numbers	14 (3.9%)
Speed-dial	37 (10.3)
Scroll through contact list/saved numbers	26 (7.6)
Voice-dial	65 (18.2)
I don't place calls while driving	216 (62.1)
Does your employer have policies in place about cell use while driving (*n* = 64)	
Yes	19 (29.7%)
No	20 (31.3)
Don't Know	25 (39.1)
Do you feel obligated to take work-related calls when driving (*n* = 64)	
Yes	19 (29.7%)
No	45 (70.3)
Driving with minors in the car (in past month)	
Driving minors under aged 11 (*n* = 343)	
Yes	92 (26.8%)
No	251 (73.2)
Driving minors aged 12–17 (*n* = 336)	
Yes	59 (17.6%)
No	277 (82.4)
Talking while driving minors under age 11 (*n* = 91)	
Never (only emergency/911 calls)	61 (67.1%)
Rarely (less than 10% of time)	27 (29.6)
Sometimes (25% of time)	1 (1.1)
Often (50% of time)	2 (2.2)
Frequently (greater than 75% of the time)	0 (0)

(continued)

Table 3. (continued)

Driving knowledge and behavior	
	N(%)/mean(SD)
Texting while driving with minors under age 11 (*n* = 89)	
Never (only emergency/911 calls)	86 (96.7%)
Rarely (less than 10% of time)	1 (1.1)
Sometimes (25% of time)	2 (2.2)
Often (50% of time)	0 (0)
Frequently (greater than 75% of the time)	0 (0)
Talking while driving minors ages 12–17 (*n* = 59)	
Never (only emergency/911 calls)	34(57.6%)
Rarely (less than 10% of time)	23 (39.0)
Sometimes (25% of time)	2 (3.4)
Often (50% of time)	0 (0)
Frequently (greater than 75% of the time)	0 (0)
Texting while driving with minors ages 12–17 (*n* = 59)	
Never (only emergency/911 calls)	54 (91.53%)
Rarely (less than 10% of time)	3 (5.1)
Sometimes (25% of time)	2 (3.4)
Often (50% of time)	0 (0)
Frequently (greater than 75% of the time)	0 (0)
Relationship with minors under age 11 (*n* = 89)	
Parent	2 (2.22%)
Grandparent	79 (87.8)
Other relative	3 (3.3)
Caregiver	1 (1.1)
Family friend	4 (4.4)
Relationship with minors ages 12–17 (*n* = 59)	
Parent	2 (3.4%)
Grandparent	44 (74.6)
Other relative	6 (10.2)
Caregiver	0 (0)
Family friend	7 (11.9)
Perceived abilities	
Ability to drive safely (*n* = 346)	
Much worse than others	0 (0%)
Somewhat worse than others	1 (0.3)
About the same as others	72 (20.8)
Somewhat better than others	119 (34.3)
Much better than others	154 (44.4)
Unsure	1 (0.3)
Ability to perform other tasks while driving (*n* = 342)	
1 (poor)	36 (10.5%)
2	40 (11.7)
3 (fair)	136 (39.8)
4	106 (31.0)
5 (excellent)	24 (7.0)
To what extent are you capable of:	
Driving safely while using handheld phone[a] (*n* = 344)	2.5 (1.1)
Driving safely while using handsfree phone[a] (*n* = 327)	3.6 (1.6)
Driving safely while texting[a] (*n* = 338)	1.4 (0.7)
Personal punitive history, knowledge, and attitudes	
How many times in the last 2 years have you been pulled over (*n* = 346)	
0	309 (89.31%)
1	35 (10.1)
2+	2 (0.6)

(continued)

Table 3. (continued)

Driving knowledge and behavior	N(%)/mean(SD)
Number of collisions in the last 2 years, regardless of fault (n = 346)	
0	281 (81.2%)
1	57 (16.5)
2+	8 (2.3)
Ever received ticket for phone use (n = 345)	
Yes	10 (2.9%)
No	335 (97.1)
Knowledge about state laws (n = 346)	
Not knowledgeable	6 (1.7%)
Somewhat knowledgeable	101 (29.2)
Very knowledgeable	231 (66.8)
I am not aware if my state has cell phone laws	8 (2.3)
My state does not have cell phone laws	0 (0)
Has your state cell phone law changed your driving behavior (n = 329)	
1 (not at all)	112 (34.0%)
2	21 (9.4)
3 (somewhat)	65 (19.8)
4	43 (13.1)
5 (significantly)	78 (23.7)
Which penalty would be effective in stopping this behavior? (n = 346)[b]	
Insurance would not cover crashed involving handheld cell phone use	4.1 (1.2)
Receiving a citation for handheld cell phone use would result in a point on your driving record	3.9 (1.2)

[a]1 (not capable at all) – 5 (very capable).
[b]1 (not effective) – 5 (very effective).

Table 4. Stepwise Linear Regression Table of Significant Coefficients

	B estimate	Std. error	p-value
Intercept	1.2963	0.7318	0.0776
Age	−0.0307	0.0143	0.0327
Number of years driving	0.0107	0.0120	0.3699
Gender (male)	0.1856	0.1047	0.0775
Number of days driving per week	0.0740	0.0372	0.0477
Smartphone ownership (yes)	0.6129	0.1376	0.0000
Perceived skill driving using phone hands-free	0.1880	0.0450	0.0000
Employed (vs. not employed)	0.7562	0.1829	0.0000
Self-employed (vs. not employed)	0.2982	0.1396	0.0335

while driving, all these drivers reported that they changed their behavior or stopped using their cell phone after the citation. Nearly 96% of participants reported feeling somewhat or very familiar with their state laws regarding cell phone calls, and 56% stated that the cell phone laws had changed their driving behaviors.

Participants were asked about hypothetical penalties for DD. 70% reported that if insurance companies did not cover crashes related to handheld cell phone use, it would be effective in changing their behavior. Similarly, 65% reported that the penalty of points on the driver's license for use of a handheld phone would be effective in changing their behavior.

DDS Outcomes and Linear Regression Models

The DDS was used to determine predictive variables for DD among the participants. Though the 11-item DDS had a possible score range of 0 to 44, the actual score ranged from 0 to 22 points among participants. The mean was about 3 points.

For all DDS questions, "never" was the most frequently selected answer. There was a high correlation between questions; participants who answered any one question with a non-zero response (endorsing any distracted behavior) were more likely to answer other questions with a non-zero response. The most commonly reported DD behaviors were questions DDS2: use of hands-free phone (51%), and DDS8: use of navigation tools (61%).

The linear regression model contained the following significant predictors (Table 4): age ($p = 0.0327$), days driving per week ($p = 0.0477$), smartphone ownership ($p < 0.0001$), perceived hands-free skill ($p < 0.0001$), and whether the driver is employed ($p < 0.0001$) or self-employed ($p = 0.0335$). Measures which did not contribute and were excluded from the final linear regression model included ethnicity, income, time spent driving per day, perceived ability to multitask while driving, perceived ability to drive safely compared to other drivers, and perceived skill at driving safely while using a handheld phone or texting.

A secondary analysis was done to include participants' medical history. Medical conditions were not significant predictors of a higher DDS score. Given this finding and a low response rate for these questions, medical conditions were also not included in the final regression model.

Discussion of Results

Despite the common perception that DD is only a problem among teens and younger adults, older adults were also found to engage in DD in this study, and interventions should include this population. Overall, cell phone use while driving was common in our study participants, as nearly 60% use their phone in some way while driving, including answering calls, making calls, and texting. Over 80% of respondents own a smartphone, which was associated with self-reported DD. Younger age, driving more miles, perceived hands-free skill, and being employed or self-employed were also associated with DD behaviors. These factors associated with higher DDS scores and riskier behaviors can inform whom to target with future interventions.

When compared with participants in the partner studies of college students (*25*) and middle-aged adults (*17*), older adults reported taking greater precautions and less frequent cell phone use while driving. Yet, most of the older drivers sampled in this study still reported DD due to cell phone use. They also reported high rates of concurrent medical conditions, including conditions that impact baseline driving abilities, though this was not significantly associated with DD behavior.

Consistent with prior studies of other age groups (*26*), the older adults reported high levels of confidence in their own driving abilities. In addition to reporting being better drivers than their peers, many older adult participants perceived their own driving ability as good or excellent even while driving distracted; this suggests that they fail to recognize the impact of DD on their driving safety. Though not directly measured, many of the participants commented on the use of functions in their car to assist with making or receiving calls, including pressing buttons on the display or steering wheel, looking at the display to see caller ID, scrolling contacts, or using navigation assistance. Though users often don't recognize these activities as DD, each require cognitive distraction, and many require the driver to take their eyes off the road. The use of navigation devices, even when pre-programmed, requires at least inattention from the task of driving, and may involve manual distraction as well (for example, when users pick up their phones to look at the screen) (*13–15*).

The survey additionally explored two specific behaviors: driving with minors, and work-related DD. Driving with minors in the car not only contributes to distraction (*27*, *28*), and models bad driving behavior for the minors (*29*, *30*), but the increased risk of crash from any DD is also putting the minor passengers at risk of injury. Nearly one in twelve participants reported using their phone while driving with younger children (<11 years old), usually their grandchildren. This unique group could benefit from targeted interventions; parents should also be encouraged to discuss safe driving with their family caretakers.

Another concerning finding was the frequency of work-related DD. 30% of employed (including self-employed) participants reported feeling obligated to take work-related calls while driving. The results from the final stepwise linear regression model showed that being self-employed or employed was a significant predictor of DD when compared to those who are not employed. This finding suggests that work needs to be done with employers, and those who are self-employed also need to be targeted and encouraged to develop clear and safe cell phone policies.

A variety of possible interventions were explored in this survey to gauge older adults' receptivity and perceived effectiveness of the interventions. The responses lend support for state laws that target driving behaviors, as over half of participants reported changing their behavior in response to laws. Respondents who previously received a citation for DD unanimously reported changing their behavior to follow the laws (i.e., hands-free), and many stopped using their phone while driving completely, though this was only a small subsample of our population (10 respondents). In terms of possible future interventions, approximately 90% of respondents stated that insurance policy changes or points on their license would at least somewhat change their behavior. Each of these potential interventions may provide an avenue to reduce DD among older adults, whether through stricter laws, increased enforcement, or punitive measures related to insurance.

In addition to these punitive measures, there are other possible mitigating strategies for behavior change. Existing education programs are offered to increase awareness about the dangers of DD. One example is the *Just Drive* class offered through the TREDS program at the University of California, San Diego. The *Just Drive* class encourages participants to be safe and responsible drivers, and targets employees. This curriculum could be modified and target other populations of older drivers, including those who are self-employed, to develop a personal culture of safety. Finally, healthcare providers could play a role; not only can providers assess risk related to concurrent medical conditions, but utilization of a simplified DDS may help identify distracted drivers who can benefit from intervention.

The authors acknowledge some limitations of this study. The study respondents were primarily White non-Hispanic, female, and in a higher income bracket than what would be representative of California. Most participants completed the survey online, which indicates the sample population may be skewed to those who have access to computers, are tech-savvy, and more likely to have smartphones. Future studies should target a more diverse population and offer materials in other languages. To target a more driving-active population, the authors chose to limit eligibility to older adults who use their cellphone at least once a week, and who drive at least once per week. Previous studies have suggested that infrequent older drivers may have higher crash risk (*31*); future studies may choose to further study this other high-risk population.

The survey was also self-reported, which may result in under-reporting of undesirable behaviors (i.e., social desirability bias), and participants were likely reluctant to admit that they use electronics while driving. The completely anonymous design of the data collection may limit some of this bias, as there was no identifying information collected. Although large side-of-road observational studies (*32*, *33*), and naturalistic driving studies using in-car monitoring systems (*34–36*) can objectively examine DD behaviors and avoid this bias, surveys such as done in this study that rely on self-reported data are less expensive and allow for large sample sizes. Additionally, this survey format allows for exploration of older adult drivers' perceptions of their own driving behavior. Future studies may consider using real-time video recordings of drivers, with concurrent use of self-reported behavior surveys; this dual technology and survey-based study could be used to validate the DDS. Finally, the results were not normally distributed because many participants reported that they never used any electronics while driving. As a cross-sectional study, variables that were associated with higher scores (i.e., predictors) were identified but conclusions about causality cannot be made.

Conclusions

With cell phones nearly ubiquitous in the U.S., prevention of cell phone use while driving is a worrisome public health and safety issue. Older drivers are a growing population that is engaging in high-risk behavior, even with medical conditions and with minors in the car. Older drivers also overestimate their driving skills and their ability to drive well while distracted. Interventions such as education, employer programs, citations, points on their driver's license, and insurance changes with DD-related crashes may mitigate these behaviors, but more studies are needed. Routine screening and education of the public may also help reduce the number of distracted drivers. These results can guide program and policy changes that could improve safety on California's roads and across the nation.

Author Contributions

The authors confirm contribution to the paper as follows: study conception and design: Hill, Rybar, Jahna; data collection: Hill, Alwahab, Chukwueke, Engler; analysis and interpretation of results: Hill, Baird, Engelberg, Larocca; draft manuscript preparation: Hill, Baird, Engler, Jahns, Rybar, Engelberg. All authors reviewed the results and approved the final version of the manuscript.

References

1. *Traffic Safety Facts 2015: A Compilation of Motor Vehicle Crash Data from the Fatality Analysis Reporting System and the General Estimates System*. Publication DOT HS 812 384. NHTSA, U.S. Department of Transportation, Washington, D.C., 2017.
2. Ferdinand, A. O., and N. Menachemi. Associations Between Driving Performance and Engaging in Secondary Tasks: A Systematic Review. *American Journal of Public Health*, Vol. 104, No. 3, 2014, pp. e39–e48. http://doi.org/10.2105/ajph.2013.301750.
3. *Mobile Fact Sheet*. Pew Research Center, January 12, 2017. http://www.pewinternet.org/fact-sheet/mobile/. Accessed June 11, 2017.
4. Mobile Device Use While Driving–United States and Seven European Countries, 2011. *Morbidity and Mortality Weekly Report*. Vol. 62, No. 10, 2013, pp. 177–182.
5. Pickrell, T. M., R. Li, and S. KC. *Driver Electronic Device Use in 2015*. Publication DOT HS 812 326. NHTSA, U.S. Department of Transportation, Washington, D.C., 2016.
6. McEvoy, S. P., M. R. Stevenson, A. T. McCartt, M. Woodward, C. Haworth, P. Palamara, and R. Cercarelli. Role of Mobile Phones in Motor Vehicle Crashes Resulting in Hospital Attendance: A Case-crossover Study. *BMJ*. Vol. 331, No. 7514, 2005, pp. 428. http://doi.org/10.1136/bmj.38537.397512.55.
7. Redelmeier, D. A., and R. J. Tibshirani. Association Between Cellular-telephone Calls and Motor Vehicle Collisions. *New England Journal of Medicine*. Vol. 336, No. 7, 1997, pp. 453–458. http://doi.org/10.1056/nejm199702133360701.
8. Braitman, K. A., and A. T. McCartt. National Reported Patterns of Driver Cell Phone Use in the United States. *Traffic Injury*

Prevention. Vol. 11, No. 6, 2010, pp. 543–548. http://doi.org/10.1080/15389588.2010.504247.

9. McEvoy, S. P., M. R. Stevenson, and M. Woodward. The Contribution of Passengers Versus Mobile Phone Use to Motor Vehicle Crashes Resulting in Hospital Attendance by the Driver. *Accident Analysis and Prevention*, Vol. 39, 2007, pp. 1170–1176. http://doi.org.10.1016/j.aap.2007.03.004.
10. Ishigami, Y., and R. M. Klein. Is a Hands-free Phone Safer Than a Handheld Phone? *Journal of Safety Research*, Vol. 40, No. 2, 2009, pp. 157–164. http://doi.org/10/1016/j.jsr.2009.02.006.
11. Leung, S., R. J. Croft, M. L. Jackson, M. E. Howard, and R. J. McKenzie. A Comparison of the Effect of Mobile Phone Use and Alcohol Consumption on Driving Simulation Performance. *Traffic Injury Prevention*, Vol. 13, No. 6, 2012, pp. 566–574. http://doi.org/10.1080/15389588.2012.683118.
12. Strayer, D. L., F. A. Drews, and D. J. Crouch. A Comparison of the Cell Phone Driver and The Drunk Driver. *Human Factors: The Journal of the Human Factors and Ergonomics Society*, Vol. 48, No. 2, 2006, pp. 381–391. http://doi.org/10.1518/001872006777724471.
13. Knapper, A., N. Van Nes, M. Christoph, M. Hagenzieker, and K. Brookhuis. The Use of Navigation Systems in Naturalistic Driving. *Traffic Injury Prevention*, Vol. 17, No. 3, 2016, pp. 264–270. https://doi.org/10.1080/15389588.2015.1077384.
14. Stimpson, J. P., F. A. Wilson, and R. L. Muelleman. Fatalities of Pedestrians, Bicycle Riders, and Motorists Due to Distracted Driving Motor Vehicle Crashes in the U.S., 2005-2010. *Public Health Reports*, Vol. 128, No. 6, 2013, pp. 436–442. https://doi.org/10.1177/003335491312800603.
15. Strayer, D. L., J. M. Cooper, J. Turrill, J. R. Coleman, and R. J. Hopman. Talking to Your Car Can Drive You to Distraction. *Cognitive Research: Principles and Implications*, Vol. 1, No. 1, 2016, pp. 16. https://doi.org/10.1186/s41235-016-0018-3.
16. Tison, J., N. Chaudhary, and L. Cosgrove. *National Phone Survey on Distracted Driving Attitudes and Behaviors*. Publication DOT HS 811 555. NHTSA, U.S. Department of Transportation, Washington, D.C., 2011.
17. Engelberg, J., L. Hill, J. Rybar, and T. Styer. Distracted Driving Behaviors Related to Cell Phone Use among Middle-Aged Adults. *Journal of Transport and Health*. Vol. 2, No. 3, 2015, pp. 434–440. http://doi.org/10.1016/j/jth.2015.05.002.
18. Betz, M. E., and S. R. Lowenstein. Driving Patterns of Older Adults: Results from the Second Injury Control and Risk Survey. *Journal of the American Geriatrics Society*, Vol. 58, No. 10, 2010, pp.1931–1935. http://doi.org/10.1111/j.1532-5415/2010/03010.x.
19. Dawson, J., E. Y. Uc, S. Anderson, A. M. Johnson, and M. Rizzo. Neuropsychological Predictors of Driving Errors in Older Adults. *Journal of the American Geriatrics Society*, Vol. 58, No. 6, 2010, pp. 1090–1096. http://doi.org/10.1111/j.1532-5415.2010.02872.x.
20. Strayer, D. L., and F. A. Drews. Profiles in Driver Distraction: Effects of Cell Phone Conversations on Younger and Older Drivers. *Human Factors*, Vol. 46, No. 4, 2004, pp. 640–649. http://doi.org/10.1518/fjes.46.4.640.56806.
21. Thompson, K. R., A. M. Johnson, J. L. Emerson, J. D. Dawson, E. R. Boer, and M. Rizzo. Distracted Driving in Elderly and Middle-aged Drivers. *Accident Analysis & Prevention*, Vol. 45, 2013, pp. 711–717. http://doi.org/10/1016/j.aap.2011.09.040.
22. Braitman, K., and A. McCartt. National Reported Patterns of Driver Cell Phone Use in the United States. *Traffic Injury Prevention*, Vol. 11, No. 6, 2010, pp. 543–548. http://doi.org/10.1080/15389588.2010.504247.
23. Donmez, B., L. N. Boyle, and J. D. Lee. The Impact of Distraction Mitigation Strategies on Driving Performance. *Human Factors*, Vol. 48, No. 4, 2006, pp. 485–804. http://doi.org/10.1518/001872006779166415.
24. Eby, D. W., L. J. Molnar, J. T. Shope, J. M. Vivoda, and T. A. Fordye. Improving Older Driver Knowledge and Self-awareness Through Self-assessment: The Driving Decisions Workbook. *Journal of Safety Research*, Vol. 34, No. 3, 2003, pp. 371–381. http://doi.org/10.1016/j.jsr.2003.09.006.
25. Hill, L., J. Rybar, T. Styer, E. Fram, G. Merchant, and A. Eastman. Distracted Driving Prevalence and Attitudes in College and University Students. *Traffic Injury Prevention*, Vol. 16, No. 4, 2015, pp. 362–367. http://doi.org/10.1080/15389588.2014.949340.1080/15389588.2014.949340.
26. Allstate Insurance Company. *New Allstate Survey Shows Americans Think They Are Great Drivers – Habits Tell a Different Story*. PRNewswire, August, 2011. https://www.prnewswire.com/news-releases/new-allstate-survey-shows-americans-think-they-are-great-drivers–habits-tell-a-different-story-126563103.html. Accessed November 2, 2017.
27. Roney, L., P. Violano, G. Klaus, R. Lofthouse, and J. Dziura. Distracted Driving Behaviors of Adults While Children are in the Car. *Journal of Trauma and Acute Care Surgery*, Vol. 75, No. 4, Suppl 3, 2013, pp. S290–S295. http://doi.org/10/1097/TA.0b013e3182924200.
28. Massey, K., S. Kant, P. Violano, L. Roney, W. King, W. Justice, K. Mcfalls, and K. Monroe. Evaluating Distracted Driving Behaviors in Parents of Children in Suburban and Rural Areas of Alabama. *Journal of Trauma and Acute Care Surgery*, Vol. 80, No. 4, Suppl 1, 2016, pp. S44–S47. https://doi.org/10.1097/TA.0000000000001181.
29. Bianchi, A., and H. Summala. The "Genetics" of Driving Behavior: Parents' Driving Style Predicts Their Children's Driving Style. *Accident; Analysis and Prevention*, Vol. 36, No. 4, 2004, pp. 655–659. https://doi.org/10.1016/S0001-4575(03)00087-3.
30. Taubman-Ben-Ari, O., O. Musicant, T. Lotan, and H. Farah. The Contribution of Parents' Driving Behavior, Family Climate for Road Safety, and Parent-targeted Intervention to Young Male Driving Behavior. *Accident; Analysis and Prevention*, Vol. 72, 2014, pp. 296–301. https://doi.org/10.1016/j.aap.2014.07.010.
31. Langford, J., J. L. Charlton, S. Koppel, A. Myers, H. Tuokko, S. Marshall, M. Man-Son-Hing, P. Darzins, M. Di Stefano, and W. Macdonald. Findings from the Candrive/Ozcandrive Study: Low Mileage Older Drivers, Crash Risk and Reduced Fitness to Drive. *Accident Analysis & Prevention*, Vol. 61, 2013, pp. 304–310. https://doi.org/10.1016/j.aap.2013.02.006.
32. Torkamannejad Sabzevari, J., A. R. Nabipour, N. Khanjani, A. Molaei Tajkooh, and M. J. M. Sullman. An Observational Study of Secondary Task Engagement While Driving on Urban Streets in Iranian Safe Communities. *Accident; Analysis and Prevention*, Vol. 96, 2016, pp. 56–63. https://doi.org/10.1016/j.aap.2016.07.020.
33. *Distracted Driving in Washington State, 2016: An Observational Survey*. Washington Traffic Safety Commission, 2017. http://wtsc.wa.gov/wp-content/uploads/2017/01/2016-DD-REPORT.pdf.

34. Marshall, S. C., M. Man-Son-Hing, J. Charlton, L. J. Molnar, S. Koppel, and D. W. Eby. The Candrive/Ozcandrive Prospective Older Driver Study: Methodology and Early Study Findings. *Accident Analysis & Prevention*, Vol. 61, 2013, pp. 233–235. https://doi.org/10.1016/j.aap.2013.07.007.
35. Li, G., D. W. Eby, R. Santos, T. J. Mielenz, L. J. Molnar, D. Strogatz, M. E. Betz, C. DiGuiseppi, L. H. Ryan, V. Jones, and S. I. Pitts. Longitudinal Research on Aging Drivers (LongROAD): Study Design and Methods. *Injury Epidemiology*, Vol. 4, No. 1, 2017, pp. 22. https://doi.org/10.1186/s40621-017-0121-z.
36. Dingus, T. A., S. G. Klauer, V. L. Neale, A. Petersen, S. Lee, J. D. Sudweeks, M. A. Perez, J. Hankey, D. J. Ramsey, and S. Gupta. *The 100-Car Naturalistic Driving Study, Phase II – Results of the 100-Car Field Experiment*. Publication DOT HS 810 593. NHTSA, U.S. Department of Transportation, Washington, D.C., 2006.

The Standing Committee on Safe Mobility of Older Persons (ANB60) peer-reviewed this paper (18-02049).

TRR
JOURNAL OF THE TRANSPORTATION RESEARCH BOARD

Article

Transportation Research Record
2018, Vol. 2672(33) 89–100

DOI: 10.1177/0361198118801356
journals.sagepub.com/home/trr
SAGE

Cognitive Change and Driving Behavior among Older Drivers

Laura Fraade-Blanar[1] **and James P. Smith**[2]

Abstract

Per vehicle miles traveled, older adults have a high fatal crash rate. One factor is dementia. This study aims to assess how differences in cognition affect driving behavior among older drivers. We analyzed data from the Health and Retirement Study. Our study used cognition, demographics, and driving behavior from 2006 to 2014 for respondents aged 65 and above. Three levels of driving behavior were measured: whether the individual could drive, whether they had driven in the past month, and whether they drove long distances. Cognitive function was measured through the Telephone Interview for Cognitive Status. Additionally, individuals were coded as having no diagnosis of dementia, a diagnosis within 2 years, or a diagnosis more than 2 years previously. We estimated the likelihood of each driving behavior in association with cognition using a modified Poisson regression model for binary outcomes. Among respondents (N = 16,061), 79% could drive. Of these, 93% had driven in the past month, and of these, 64% drove long distances. Compared with no impairment, mild impairment was associated with a significant 12% decrease in probability of being able to drive, an 8% decrease in driving within the past month, and a 24% decrease in driving long distances. The decrease was larger among those with severe impairment. Results were in a similar direction and strength comparing individuals without dementia with individuals 0 to 2 years after diagnosis, and to more than 2 years after diagnosis. A strong positive association exists between lower cognition and lower driving exposure.

Older adults are a large presence on the roads in the United States: 40.1 million adults aged 65 and above had an active driver's license in 2015, representing a 33% increase from 2005 (*1*). By 2050, one out of every four drivers on the road will be aged 65 or older (*2*). Driving is beneficial to older adults (*3, 4*), with cessation linked to depression, isolation, entry into nursing homes, and greater burden for the care provider (*2, 3, 5, 6*). However, a crash may be devastating for older adults and endanger other road users.

Per vehicle miles traveled, adults aged 65 and above have a high fatal crash rate compared with other age groups (*7*). One contributing factor is cognitive decline or dementia (*3, 5, 8–10*). Cognitive impairment is particularly important because it affects two of the main means of preserving older driver safety: self-monitoring and self-regulation. Self-monitoring refers to awareness of one's driving ability. Self-regulation refers to the ability to set and enforce safe driving habits (*2, 8, 10–12*).

Research has shown that dementia can impair driving skill via various pathways: decreased perception of hazards and processing of visual cues, diminished attention to the driving task, impaired decision making, and slower reaction times to on-road situations (*3, 5, 8–10, 13*). Older drivers with impaired cognitive function may become lost more frequently, have difficulty at intersections, and may stray from marked lanes and customary routes (*9, 14–17*).

Generally, older adults with mild dementia are considered able to drive by the National Highway Traffic Safety Administration, the American Academy of Neurology, the American Geriatrics Society, and state departments of motor vehicles (*2, 18, 19*). Many continue to drive, albeit with lower mileage (*17, 20, 21*). Indeed, in two small cross-sectional studies, self-reported mileage was 15% to 42% lower among those with a CDR (clinical dementia rating) of 0.5 (very mild dementia) and 46% to 64% lower among those with a CDR of 1.0 (mild dementia) compared with a CDR of 0 (normal cognition) (*13, 17*). A study of 18 dementia patients and their age-matched controls using caregiver reports found

[1]Center for the Study of Aging, Labor & Population, RAND Corporation, Santa Monica, CA
[2]Chair in Labor Markets and Demographic Studies, RAND Corporation, Santa Monica, CA

Corresponding Author:
Address correspondence to Laura Fraade-Blanar: lblanar@rand.org

respondents with dementia drove 45% fewer miles compared with those without dementia (*22*).

However, self-reported mileage and driving behavior can be inaccurate (*23–26*). This may be particularly true for drivers with low mileage, a common feature of older driver behavior (*27*). In addition, not all mileage is equal: different roadways, geographic areas, times of day, days of the week, certain weather, and certain vehicle types can make differing demands on the driver and produce different crash risks (*15*, *28*). Exposure to these risks may vary with cognition; a study showed drivers in early stage dementia had fewer unique destinations, and more frequently drove during daylight, within 5 mi of home, and slower than surrounding traffic (*29*). Consequently, there remains a lack of understanding around how driving behavior changes quantitatively and qualitatively over time in association with cognition (*30*). This study aims to assess how differences in cognition affect measures of driving behavior among older drivers.

Methods

Participants

This study was a retrospective secondary analysis of longitudinal survey data from the Health and Retirement Survey (HRS). The National Institute on Aging (grant number NIA U01AG009740) sponsors the HRS through the University of Michigan. The HRS's content and methodologies have been described elsewhere (*31–34*). Briefly, data collection began in 1992, subsequently adding new content and expanding cohorts(*34*). Participants are interviewed every 2 years. Interviews can be conducted in-person or via phone. No differences have been found in cognitive testing performance between these modes (*35*). The HRS gathers data on health, employment, economic resources, and demographic variables (*36*). The survey maintains a response rate over 75% in all groups except Hispanics (*37*). Participants are Americans aged 50 years and older.

Measures

The HRS asks three questions specific to driving: "Are you able to drive?" "Have you driven a car in the past month?" and "Do you limit your driving to nearby places, or do you also drive on longer trips?" Each follow-up question is only asked if the previous one is answered affirmatively. Individuals were asked driving exposure questions anew during each study panel.

Further questions delve into driving circumstances. All respondents are asked about driving resources, specifically, "Do you have a car available to use when you need one?" Questions into driving behavior include, "Because of a health or memory problem, do you have any difficulty using a map to figure out how to get around in a strange place?" and, "When you ride in a car, how often do you wear your seat belt? Is it all of the time, most of the time, sometimes, rarely, or never?"(*36*).

We used two measures of cognition. First, cognitive functioning was measured through a combined score of memory and mental status, as adapted from the Telephone Interview for Cognitive Status (TICS) (*38–41*). To increase interpretability, individual cut-offs were established as per previous research, (*42*, *43*) at lower than two standard deviations below the mean (severely cognitively impaired), lower than one deviation below (mildly impaired), and higher than one deviation below the mean (no impairment). Analyses using this measure included data from 2006 to 2012, the first panel in which all three driving measures were obtained, to the most recent panel in which TICS is available.

The second measure involved diagnosed dementia. Starting in 2012, individuals were coded as reporting a diagnosis of dementia, including Alzheimer's disease, in the previous 2 years or ever. Measures were generated of individuals who had received a diagnosis of dementia in the last 2 years, more than 2 years ago, or never having had a diagnosis of dementia. This analysis used data from 2012 and 2014, the two panels with both dementia diagnosis information and driving exposure measures.

Statistical models included potential confounders related to health and demographics. With regard to health, self-reported vision was grouped into poor to fair and good to excellent (*36*). To adjust for comorbidities, individuals were categorized as having zero, one to three, or four or more of the following conditions: high blood pressure or hypertension; diabetes or high blood sugar; cancer (except skin cancer); chronic lung disease (except asthma); heart problems; stroke; emotional, nervous, or psychiatric problems; and arthritis or rheumatism (*34*). Spousal health conditions were grouped in the same way. Spousal health conditions were included to explore the availability of alternative drivers. We hypothesized that a healthy spouse may allow respondents with impaired cognition to restrict or cease driving with greater ease.

Demographics included gender, age (categorized as 65 to 74, 75 to 84, and 85 and above), wealth, and education. Wealth, defined as net value or the sum of all wealth components, including second houses, minus any debt (*34*), was quartiled. Education was grouped into having less than a high school education, one or more years of high school, or more than a high school education.

Additional measures aimed to capture access to transportation resources beyond driving (e.g., public transit or other drivers in the living situation). Individuals were categorized as rural, suburban, or urban according to the Beale Rural–Urban Continuum Code (*44*). Additional measures were included if the individual had a spouse or lived with a partner, if there was a child living at home or within 10 mi of the respondent, and if the individual was living in a nursing home or other healthcare facility.

Inclusion Criteria

We included participants aged 65 and above, as driving exposure questions are only asked of these individuals. Respondent reports that they were unable to drive, had not driven in the past month, or did not drive on longer trips did not result in exclusion from current or subsequent panels. Individuals were followed as long as they were retained in the HRS and responded to survey questions. Individual who were legally blind were removed from the sample ($N = 325$).

Analysis

We estimated the likelihood associated with each driving behavior in association with each cognitive measure using population average estimates, generated from a generalized estimating equation model (*45*). Within this framework we used a modified Poisson distribution for binary outcomes (*46*) with robust standard errors. To account for the correlation between repeated observations from the same respondent, we clustered data on the individual level (*45*). Pearson's goodness of fit tests were not significant, suggesting reasonable model fit for Poisson models. We also examined a random effects model, which yielded very similar results.

Cognitive measures were modeled as a dummy variable with no impairment/no dementia as the reference categories. All models were adjusted for eyesight, health, spousal health, gender, age, wealth, education, urbanicity, presence of a spouse or live-in partner, living in a nursing home, and a child at home or living within 10 mi. No model adjusted for previous cognitive measures. A separate model involving only those with a spouse was created so that the measure of spousal health could be included.

A secondary analysis among individuals who were severely impaired or were more than 2 years post-dementia diagnosis examined the likelihood of driving without any attenuation in exposure (being able to drive, driving at least once a month and driving long trips). The analysis used the same analytic model.

Additionally, models were created for two types of outcome measures: 1. Behaviors around map reading ability and seat-belt wearing, and 2. The driving resource measure, concerning access to a car. Within these models, cognitive measures were again the primary predictor of interest and all models were adjusted for eyesight, health, spousal health, gender, age, wealth, education, urbanicity, presence of a spouse or live-in partner, living in a nursing home, and a child at home or living within 10 mi.

Sub-analyses were used to further explore the data. To minimize missing data, if an individual is unable to participate in the survey, the HRS attempts to use a proxy, generally a spouse or child (*34*, *40*). Data contributed by proxies includes dementia diagnosis. To explore possible bias within the dementia diagnosis model, we conducted a sub-analysis excluding all proxy response-based surveys. HRS does not use proxy responses for TICS.

A second sub-analysis adjusted for HRS panel year to account for chronological differences. Although we explored a variety of models for panel year, there was no evidence of any relationship between year, driving behavior, and cognition, nor did we find any trend.

Analyses were performed using STATA version 13.1 (StataCorp, College Station, TX).

Results

Participant Demographics

Between 2 and 10 years' worth of data was used from individual participants ($N = 16{,}061$) from the HRS. Within this study, 56% of respondents provided data from five panels (2006 to 2014). Between 9% and 12% of respondents provided data for each other possibility (data from one panel, two panels, three panels, or four panels). This combines to 54,441 panels overall. In 79% of interviews, respondents stated they could drive. Of these, 93% had driven in the past month, and of these, 64% drove long distances as opposed to short distances (Figure 1). As Table 1 shows, individuals who reported being able to drive generally had less cognitive impairment, no dementia, were younger, more frequently had a live-in partner or spouse, more commonly lived in an urban setting, had more education, more wealth, were more commonly male, had better eyesight, and had fewer co-morbid conditions compared with individuals unable to drive.

Cognition and Driving Behavior

Among respondents, compared with no impairment, mild impairment was associated with a 12% lower probability of reporting being able to drive (p-value < 0.01), an 8% lower probability of having driven within the past month (p-value < 0.01) among those able to drive, and a 24% lower probability of driving long distances (p-value < 0.01) among those who drove in the past month (Figure 2). The reduction in probability was larger among those with severe impairment.

Among potential confounders, living in a rural or suburban area compared with an urban area, having a

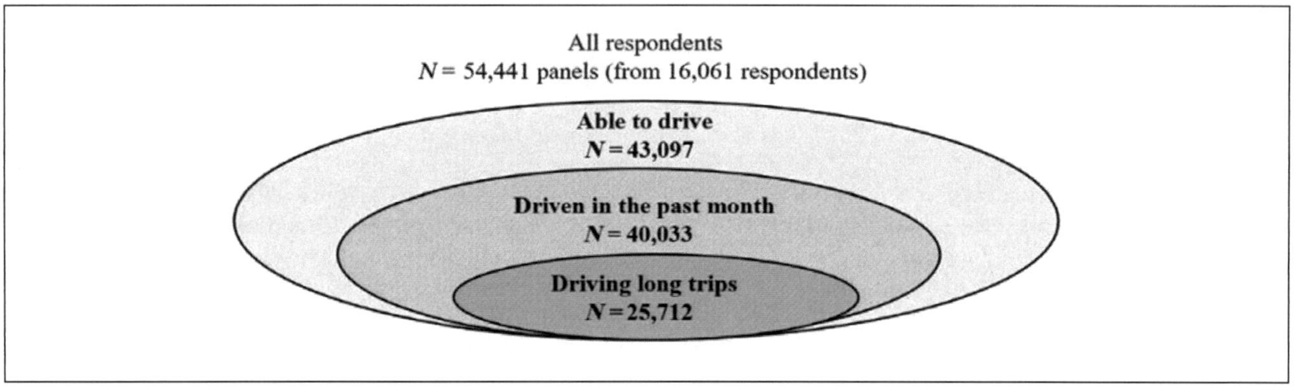

Figure 1. Responses to questions around driver behavior.

spouse or live-in partner, higher levels of education, higher levels of wealth, and having good to excellent eyesight was associated with a greater likelihood of reporting being able to drive (Table 2). Being female, living in a nursing home, having children cohabiting or living nearby, and higher numbers of comorbidities were associated with a lower likelihood of ability to drive (p-values < 0.01). These associations did not change significance or strength of association when modeled with driving at least once a month or driving long distances as the outcomes.

In a sub-analysis of individuals with spouses, the count of spousal comorbidities was not statistically significant. As an exception, analyses with driving once a month and long distances as outcomes showed a significant association among those with a spouse with four or more health conditions compared with no health conditions. Adjusted relative risks (aRRs) were alternatively associated with a higher likelihood of driving at least once a month (aRR = 1.02, p-value = 0.04) and a lower likelihood of driving long distances (aRR = 0.93, p-value < 0.01). Inclusion of spousal health status did not change the relationship between cognition and driving behavior, and no interaction was present.

Dementia Diagnosis and Driving Behavior

There was a lower likelihood of reporting each driving behavior among those with a dementia diagnosis compared with those without a dementia diagnosis. Among respondents, compared with no dementia, 0 to 2 years post-diagnosis was associated with a 44% lower probability of reporting being able to drive (p-value < 0.01), a 14% lower probability of driving within the past month (p-value < 0.01) among those able to drive, and a 28% lower probability of driving long distances (p-value $<$ 0.01) among those who had driven in the past month. Results were in a similar direction and strength comparing individuals not diagnosed with dementia with individuals more than 2 years after diagnosis (Figure 3).

In a sub-analysis of individuals with spouses, the count of spousal comorbidities was statistically significant in association with ability to drive but not with other driving exposure measures. Having a spouse with one to three comorbidities was associated with an aRR of 1.04 (p-value < 0.01) and four or more conditions with a relative risk of 1.04 (p-value = 0.02), compared with no comorbidities. Inclusion of spousal health status did not change the relationship between cognition and driving behavior, and no interaction was present.

In a sub-analysis around proxy interviews, a chi-squared test showed a significant relationship between dementia diagnosis and use of a proxy. Eight percent of surveys used proxy responses. Among those 0 to 2 years post-dementia diagnosis, 47% used a proxy. Among those more than 2 years post-dementia diagnosis, 65% used a proxy. When proxy responses were removed, the relationship between diagnosis and ability to drive was slightly attenuated but still significant ($p < 0.01$). There was no change in models around driving in the past month and driving long trips.

Characteristics of Drivers Who Are Severely Impaired, Are More Than 2-Years Post-Dementia Diagnosis, or Both

In 187 interviews (0.34%, data from 165 respondents), individuals stated they were currently driving long distances and were either considered severely impaired as per the TICS or were more than 2-years post-dementia diagnosis. Within this group, respondents who are more likely to report driving without attenuation (i.e., able to drive, driving at least once a month, and driving long distances) included those living in suburban and rural settings at that time (Table 3). Being age 85 and above, being a woman, and having a higher number of

Table 1. Sample Demographics, Divided by Self-Reported Being Able to Drive (*N* = 16,061) at First In-Study Interview

	Unable to drive		Able to drive	
	N (*N* = 3,034)	%	*N* (*N* = 13,027)	%
Cognition[a,c]				
no impairment	1,250	58	10,347	89
mild impairment	550	25	1,024	9
severe impairment	370	17	204	2
Dementia diagnosis[a,b]				
no dementia	177	85	1,817	99
0–2 years post diagnosis	20	10	10	1
3+ years post diagnosis	11	5	12	1
Age group[a]				
65–74	1,154	38	9,160	70
75–84	925	31	3,043	23
85+	952	31	809	6
Spouse or live-in partner[a]	1,140	38	8,803	73
Education[a]				
less than high school	892	29	1,135	9
at least 1 year of high school	1,510	50	6,193	48
more than high school	630	21	5,687	44
Wealth[a]				
quartile 1	1,402	46	1,865	14
quartile 2	799	26	3,063	24
quartile 3	497	16	3,871	30
quartile 4	336	11	4,228	32
Male[a]	678	22	6,137	47
Female[a]	2,356	78	6,888	53
Eyesight[a]				
fair to poor	1,446	48	2,428	19
good to excellent	1,564	52	10,588	81
Number of comorbidities[a]				
none	130	4	1,348	10
1 to 3 conditions	1,863	61	9,496	73
4+ conditions	1,041	34	2,181	17
Number of spouse's comorbidities[a,d]				
none	116	10	984	12
1 to 3 conditions	760	68	6,148	72
4+ conditions	235	21	1,372	16
Urbanicity [a]				
urban	1,572	52	5,961	46
suburban	696	23	3,086	24
rural	745	25	3,922	30
A child in house or up to 10 mi away[a]	2,053	68	7,890	66
In a nursing home[a]	463	15	35	0.3

[a]Statistically significant (*p*-value < 0.05).
[b]Only reflects data from 2012 and 2014 panels.
[c]Only reflects data from 2006 to 2012.
[d]Only reflects those with a spouse.

comorbidities was associated with a lower likelihood of reporting no attenuation in driving among individuals with severe cognitive impairment or who are three or more years post-dementia diagnosis (Table 3).

Driving Skills, Resources, and Cognition

In relation to driving skills and resources, 86% of individuals reported no problems reading maps, 97% reported access to a car, and 97% reported wearing their seat belt all the time or most of the time.

Difficulty reading a map was associated with lower cognition compared with no impairment. The aRR for mild impairment was 1.36 (p-value < 0.01). For severe impairment, the aRR was 1.90 (p-value < 0.01). Individuals with impairment reported lower access to cars. Compared with no impairment, the aRR for mild impairment was 0.97 (p-value < 0.01), and for severe

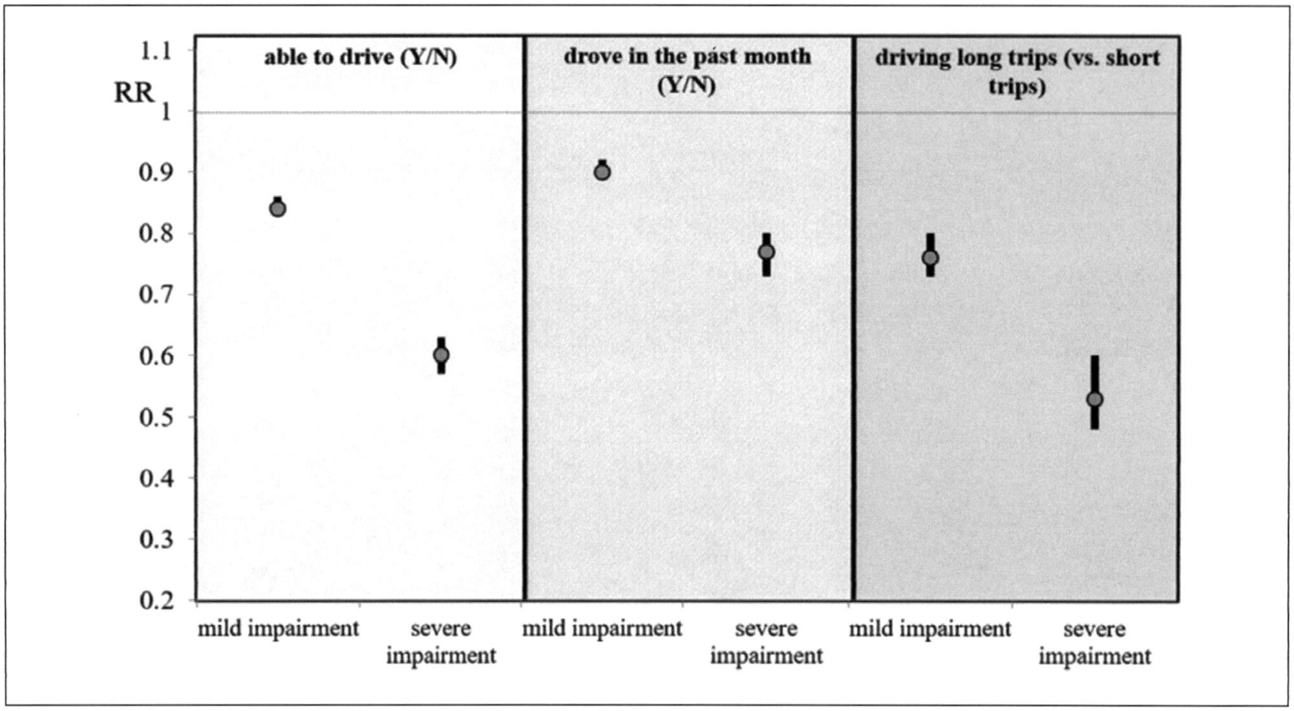

Figure 2. Adjusted relative risk (aRR)[a] and 95% confidence interval[b] of each driving behavior (reference: no impairment).
Note: [a]All analyses adjusted for age, gender, vision, wealth, education, the presence of a spouse or live-in partner, urbanicity, being in a nursing home, and comorbidities.
[b]All estimates were statistically significant.

Table 2. Adjusted Relative Risk (aRR) Associated with Reporting Being Able to Drive (95% Confidence Interval)

	aRR	95% Confidence interval[a]
No impairment (reference group)		
Mild impairment	0.88	0.87, 0.90
Severe impairment	0.66	0.63, 0.69
Urban (reference group)		
Suburban	1.04	1.03, 1.06
Rural	1.07	1.05, 1.08
Age 65–74 (reference group)		
Age 75–84	0.96	0.95, 0.96
Age 85+	0.80	0.78, 0.82
A spouse or live-in partner (no partner as reference group)	1.03	1.02, 1.04
Living in a nursing home	0.41	0.35 0.47
A child in house or up to 10 mi away	0.98	0.97, 0.99
No high school (reference group)		
At least 1 year of high school	1.20	1.15, 1.24
More than high school	1.23	1.19, 1.27
Wealth quartile 1 (reference group)		
Wealth quartile 2	1.12	1.09, 1.13
Wealth quartile 3	1.16	1.14, 1.18
Wealth quartile 4	1.18	1.16, 1.20
Female (male as reference group)	0.86	0.85, 0.87
No comorbidities		
1 to 3 conditions	0.97	0.96, 0.98
4+ conditions	0.91	0.89, 0.92
Good to excellent eyesight (fair to poor as the reference group)	1.08	1.07, 1.09

[a]All p-values < 0.01.

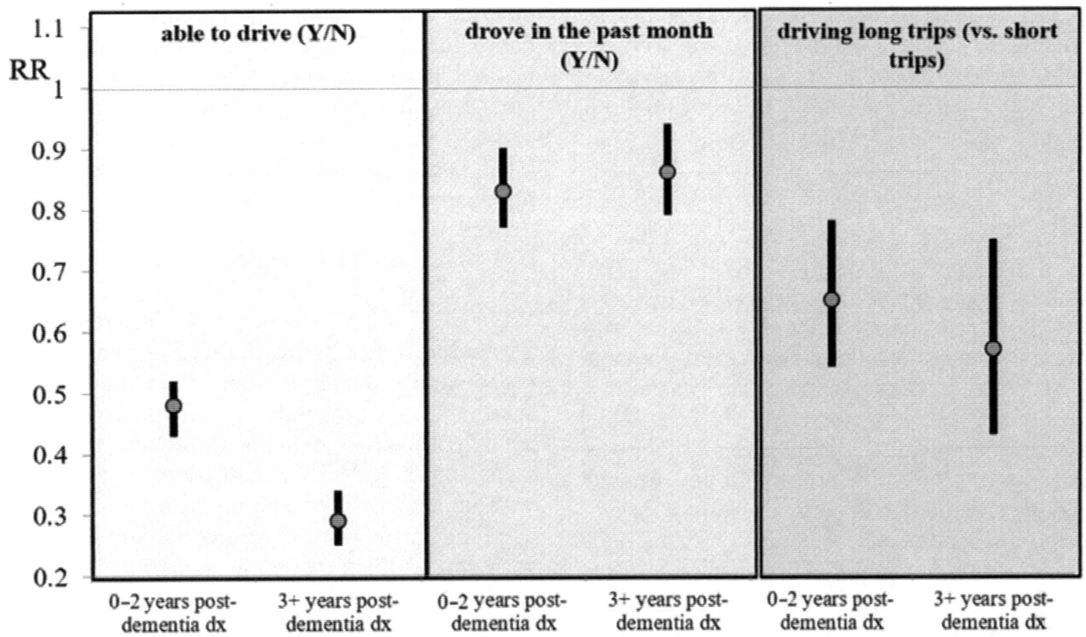

Figure 3. Adjusted relative risk (aRR)[a] and 95% confidence interval[b] of each driving behavior (reference: no dementia).
[a]All analyses adjusted for age, gender, vision, wealth, education, the presence of a spouse or live-in partner, urbanicity, being in a nursing home, and comorbidities.
[b]All estimates were statistically significant.

Table 3. Adjusted Relative Risk (aRR) Associated with Reporting No Attenuation in Driving (Being Able to Drive, Driving at Least Once a Month, and Driving Long Distances) among Individuals with Severe Cognitive Impairment or Who Are Three or More Years Post-Dementia Diagnosis

	aRR	95% confidence interval	*p*-value
Urban (reference group)			
Suburban	1.58	1.05, 2.38	0.03
Rural	1.87	1.29, 2.73	<0.01
Age 65–74 (reference group)			
Age 75–84	1.06	0.76, 1.5	0.72
Age 85+	0.42	0.26, 0.69	<0.01
A spouse or live-in partner (no partner as reference group)	1.26	0.86, 1.85	0.23
No high school (reference group)			
At least 1 year of high school	0.82	0.57, 1.17	0.27
More than high school	0.97	0.59, 1.59	0.90
Wealth quartile 1 (reference group)			
Wealth quartile 2	1.02	0.66, 1.57	0.94
Wealth quartile 3	1.53	0.97, 2.4	0.07
Wealth quartile 4	1.12	0.62, 2.02	0.71
Female (male as reference group)	0.27	0.18, 0.4	<0.01
No comorbidities			
1 to 3 conditions	0.75	0.22, 1.32	0.33
4+ conditions	0.48	0.25, 0.9	0.02
A child in house or up to 10 mi away	0.89	0.64, 1.24	0.49
Good to excellent eyesight (fair to poor as the reference group)	1.27	0.91, 1.78	0.15

impairment the aRR was 0.91 (p-value < 0.01). There was no significant relationship between seat-belt use and impairment.

The relationships between diagnosed dementia and driving skills and resources were similar in strength and direction to the above models using cognitive impairment, although in the case of map reading, the relationship was stronger. Compared with no dementia, the aRR for zero to 2-years post-diagnosis was 2.28 (p-value < 0.01), and for severe impairment the aRR was 2.57 (p-value < 0.01). There was no significant relationship between seat-belt use and dementia diagnosis.

Discussion

Overall this study showed a strong direct relationship between cognition and driving; as cognition lowered, exposure dropped. This relationship was present both when cognition was measured along a spectrum (through TICS) and by dementia diagnosis. With one exception—driving once a month as an outcome with dementia diagnosis as the cognition measure—the likelihood of a given driving behavior was closer to 1.00 for mild impairment/0 to 2 years post-dementia diagnosis than for severe impairment/more than 2 years post-diagnosis. This finding agrees with the literature, which has shown lower cognition associated with lower levels of exposure, (*13, 17, 22, 47*) but is among the first to show these measures from a longitudinal study, and to show cognition both through an incremental measurement and through dementia diagnosis.

Generally, strength and direction of association measured by TICS and years post-dementia diagnosis agreed across driving behavior, resources, and skills. The relationship between self-rated driving ability and cognition was stronger with dementia, whereas the relationships between driving in the past month, long driving trips, and cognition was stronger in the TICS-based models. This may reflect a sudden change in driving habits following a dementia diagnosis. However, among those who continue to drive post-diagnosis, habits do not change considerably. Indeed, in relation to the likelihood of driving once a month or driving long distances, more than 2 years post-dementia diagnosis did not differ significantly from 0 to 2 years post-diagnosis. Both differed significantly from no diagnosis. In comparison, the TICS measurements and association with driving behavior represent more of a gradual adjustment to a transition from normal to impaired cognition (*39*). Although TICS is not a measure of global cognition, it is arguable that the TICS-based measure reflects the association between cognition and driving behavior, and that the dementia diagnosis-based driving measure reflects the association between disease diagnosis and driving behavior. Additionally, because TICS may have a finer measurement of cognition beyond the presence or absence of a diagnosis, it may be able to suggest level of impairment and hence driving ability with more precision.

This differentiation is important as dementia may be under-screened and underdiagnosed (*48–50*). Indeed, a 2009 study estimated dementia was underdiagnosed in 50% of primary care patients aged 65 and above (*48*). Consequently, within the analyses using dementia diagnosis, exposure misclassification may result in individuals with dementia being categorized as no dementia diagnosis because they have yet to be diagnosed. Dementia diagnosis could be conceptualized within this context as a highly effective intervention to persuade potentially risky drivers to review, reduce, or cease their driving exposure.

The variation between cognition and exposure agrees with research around motor vehicle crashes. Studies from Western Australia, Quebec, and Washington State found a lower crash risk among individuals with dementia compared with those without dementia, unadjusted for exposure (*51–53*). Conversely, a small American study found a higher crash risk among those with cognitive impairment and mild Alzheimer's disease compared with those without dementia, adjusted for self-reported weekly mileage (*13*). In the first set of studies, it is likely individuals with impairment were not driving and hence not crashing.

Study results showed different driving skills (e.g., ability to read a map) with differing cognition. The relationship is more pronounced between behavioral measures and diagnosed dementia, compared with behavioral measures and impairment. Other studies showed a relationship between low seat-belt use and lower cognition or dementia (*29*), suggesting a lack of self-awareness and self-regulation ([American Geriatrics Society 2016] *54*). The relationship was not present in this study, possibly because over 97% of the sample always or almost always wore their seat belt.

Although most individuals had access to a vehicle, the relationship between having access to a vehicle and cognition shows lower access with lower cognition. This could simply relate to higher amounts of driving cessation: individuals with dementia may cease driving and hence not require access to a car. However, it could also reflect factors completely unrelated to cognition, such as access to public transit.

Although some results from the secondary analysis around spousal health were significant, the results were not qualitatively meaningful (e.g., an aRR of 1.04). Other studies found the presence of a spouse to be related to driving restrictions (*55*). Limited findings from another study showed that proximity of a child did increase the likelihood of parental driving restriction (*47*), but generally the child's role is complicated, blending concern for parents' driving abilities and fears that the parents will not respond to suggestions of restricted driving (*56*) with worries over the burden of becoming their parents' primary means of transportation (*57*).

Both measures of cognition show a small minority of individuals who are still driving despite severe impairment, or being more than 2 years post-dementia diagnosis, or both. This behavior is unrelated to the presence of a spouse or child, that is, other potential surrogate drivers. This study found driving behaviors were also not associated with wealth, which could otherwise suggest having the resources to purchase driving alternatives. Indeed, the only significant associations, besides gender and age, were comorbidities and urbanicity. This suggests individuals are driving not because they lack alternatives (e.g., taxis, family members, or, drawing on the analysis around spousal health, a healthy spouse who could drive), but rather because they believe they still have the ability to do so. The lack of other comorbidities, the less challenging driving environment presented by suburban and rural roads, or both these factors, could contribute to this self-assessment. It is also possible that respondents are inaccurately reporting their driving habits, and this inaccuracy is particularly occurring among individuals who are severely impaired or more than 2 years post-dementia diagnosis. More research is needed.

Limitations

All responses were self-reported. Even when aiming to be truthful, individuals are not always accurate in reporting their driving (*58*). This may be particularly true for individuals with cognitive impairment (*26*). Driving exposure responses were not validated, nor was data on age, spouse, living situation, and so forth. This concern is somewhat mitigated by individuals being asked about their habits rather than specific mileage. However, for the driving measures around ability to drive, and driving long distances, there is a level of subjectivity. A respondent may not have driven in years or feel comfortable driving, yet they still may believe that in an emergency they could drive. Similarly, with no definition of "long distances," respondents' interpretation of the question may differ.

The self-reported nature of responses is particularly of concern among those severely impaired or more than 2 years after a dementia diagnosis. Because of their impaired cognitive state, their responses may not reflect their habits. Indeed, it is possible that at least some of those individuals in this group who state they have no exposure attenuation per HRS measures may be inaccurately reporting their habits.

This study concomitantly includes in-person and phone-based interviews. According to the HRS codebook, Wave 1 interviews, interviews with respondents 80 years of age or older, initial interviews, and starting with the HRS 2006 wave, about half of the sample had in-person interviews, with the rest via phone (*34*). Studies have not shown differences in data quality between in-person and telephone-based surveys (*35*). Individuals from the Asset and Health Dynamics study (a sister study to the HRS) were randomly assigned to telephone or in-person surveys. Researchers found no differences in average cognitive scores based on survey mode (*35*). Additionally, papers found considerable overlap between results from TICS (conducted via phone) and Mini Mental State Exam scores and between results from TICS and in-person neurophysiological test battery (*39, 59*). However, it was possible that phone-based interviews performed differently than non-phone-based interviews.

In addition, cognitive measures in the HRS present a challenge. Around TICS, previous research showed that HRS-based cognitive tests had difficulty distinguishing mild impairment from severe or no impairment (*41*). Around dementia diagnosis, the use of proxy interviews calls the data quality into question. Encouragingly, there were only minor differences between models with and without proxy responses. Because proxy interviews yield dementia diagnosis information but not TICS scores, the TICS analyses may be missing the most severely impaired individuals. As TICS data and dementia diagnosis data only overlap for 1 year in this study, it is not possible to draw firm conclusions around their relationship.

Finally, the driving exposure measures are limited to those present in HRS. Although these measures provide an informative overview of driving habits, further details of driving behavior, such as driving at night, driving in areas the respondent is unfamiliar with, driving on highways and interstates, and driving alone, would be useful. Because the HRS lacks information on driving performance and outcomes (i.e., crashes), it is only possible to determine that the individual stated they were driving, not that this continued driving is necessarily appropriate or safe.

Future research on older adults might consider technologic approaches to measure driving habits. Such approaches could include cell phone applications, in-vehicle technology including cameras, GPS, accelerometers, and actigraphs. Data on driving habits might inform researchers and clinicians of risk and protective factors to guide decision making about driving for older adults and their families.

Conclusion

Overall this study suggested that, on an academic level, research around driving safety and cognition must account for driving exposure differences between cognitive status or risk biased results. On a clinical level, study results suggested that driving attenuation may be occurring at appropriate intervals, albeit to an incomplete degree. It also showed that, as cognition lowers, driving-related skills such as map reading also drop. Overall there is a need for more research on safety-related behavior, and objective, non-self-reported driving exposure data to facilitate our understanding of how decisions on driving habits are made.

Author Contributions

The authors confirm contribution to the paper as follows: study conception and design: LF-B, JS; data collection: LF-B; analysis and interpretation of results: LF-B; draft manuscript preparation: LF-B, JS. All authors reviewed the results and approved the final version of the manuscript.

References

1. National Highway Traffic Safety Administration. *Traffic Safety Facts 2015 Data: Older Population* [Internet], 2017. http://www-nrd.nhtsa.dot.gov/Pubs/811659.pdf. Accessed August 10, 2015.
2. American Geriatrics Society; and Pomidor, A., ed. *Clinician's Guide to Assessing and Counseling Older Driver*, 3rd ed. American Geriatrics Society, National Highway Traffic Safety Administration, Washington, D.C., 2016.
3. Martin, A., R. Marottoli, and D. O'Neill. Driving Assessment for Maintaining Mobility and Safety in Drivers with Dementia (Review). *Cochrane Database of Systematic Reviews*, No. 10, 2011.
4. Ragland, D. R., W. A. Satariano, and K. E. MacLeod. Driving Cessation and Increased Depressive Symptoms. *Journal of Gerontology*, Vol. 60, No. 3, 2005, pp. 399–403.
5. Breen, D. A., D. P. Breen, J. W. Moore, P. A. Breen, and D. O'Neill. Driving and Dementia. *British Medical Journal*, Vol. 334, No. 7608, 2007, pp. 1365–1369.
6. Freeman, E. E., S. J. Gange, B. Munoz, and S. K. West. Driving Status and Risk of Entry into Long-Term Care in Older Adults. *American Journal of Public Health*, Vol. 96, No. 7, 2006, pp. 1254–1259.
7. Insurance Institute for Highway Safety. *Fatality Facts: Older People* [Internet], 2016. http://www.iihs.org/iihs/topics/t/older-drivers/fatalityfacts/older-people. Accessed July 31, 2016.
8. Horswill, M., S. Marrington, C. McCullough, J. Wood, N. Pachana, J. McWilliam, and M. K. Raikos. The Hazard Perception Ability of Older Drivers. *The Journals of Gerontology Series B: Psychological Sciences and Social Sciences*, Vol. 63, No. 4, 2008, P212–P218.
9. Withaar, F., W. Brouwer, and A. van Zomeren. Fitness to Drive in Older Drivers with Cognitive Impairment. *Journal of the International Neuropsychological Society*, Vol. 6, No. 4, 2000, pp. 480–490.
10. Anstey, K. J., J. Wood, S. Lord, and J. G. Walker. Cognitive, Sensory and Physical Factors Enabling Driving Safety in Older Adults. *Clinical Psychology Review*, Vol. 25, No. 1, 2005, pp. 45–65.
11. Baldock, M., J. Mathias, A. McLean, and A. Berndt. Self-Regulation of Driving and Its Relationship to Driving Ability among Older Adults. *Accident Analysis & Prevention*, Vol. 38, No. 5, 2006,pp. 1038–1045.
12. Molnar, L. J., J. L. Charlton, D. W. Eby, J. Langford, S. Koppel, G. E. Kolenic, and S. Marshall. Factors Affecting Self-regulatory Driving Practices Among Older Adults. *Traffic Injury Prevention*, Vol. 15, No. 3, 2014, pp. 262–272.
13. Ott, B. R., W. C. Heindel, G. D. Papandonatos, E. K. Festa, J. D. Davis, L. A. Daiello, and J. C. Morris. A Longitudinal Study of Drivers with Alzheimer Disease. *Neurology*, Vol. 70, No. 14, 2008, p. 1109.
14. Dawson, J. D., S. W. Anderson, E. Y. Uc, E. Dastrup, and M. Rizzo. Predictors of Driving Safety in Early Alzheimer Disease. *Neurology*, Vol. 72, No. 6, 2009, pp. 521–527.
15. Wagner, J. T., R. M. Müri, T. Nef, and U. P. Mosimann. Cognition and Driving in Older Persons. *Swiss Medical Weekly*, Vol. 140, 2011.
16. Rizzo, M., D. V. McGehee, J. D. Dawson, and S. N. Anderson. Simulated Car Crashes at Intersections in Drivers with Alzheimer Disease. *Alzheimer Disease and Associated Disorders*, Vol. 15, No. 1, 2001, pp. 10–20.
17. Carr, D., J. Duchek, and J. Morris. Characteristics of Motor Vehicle Crashes of Drivers with Dementia of the Alzheimer Type. *Journal of the American Geriatrics Society*, Vol. 48, No. 1, 2000.
18. Gergerich, E. M. Reporting Policy Regarding Drivers with Dementia. *Gerontologist*, Vol. 56, No. 2, 2016, pp. 345–356.
19. Iverson, D. J., G. S. Gronseth, M. A. Reger, S. Classen, R. M. Dubinsky, and M. Rizzo. Practice Parameter Update: Evaluation and Management of Driving Risk in Dementia. *Neurology*, Vol. 74, No. 16, 2010, pp. 1316–1324.
20. Foley, D., and K. Masaki. Driving Cessation in Older Men with Incident Dementia. *Journal of the American Geriatrics Society*, Vol. 48, No. 8, 2000, pp. 928–930.
21. Lyman, J. M., G. McGwin, and R. V. Sims. Factors Related to Driving Difficulty and Habits in Older Drivers. *Accident Analysis & Prevention*, Vol. 33, No. 3, 2001, pp. 413–421.
22. Festa, E. K., B. R. Ott, K. J. Manning, J. D. Davis, and W. C. Heindel. Effect of Cognitive Status on Self-Regulatory Driving Behavior in Older Adults: An Assessment of Naturalistic Driving Using In-Car Video Recordings. *Journal of Geriatric Psychiatry and Neurology*, Vol. 26, No. 1, 2013, pp. 10–18.
23. Blanchard, R. A., A. M. Myers, and M. M. Porter. Correspondence between Self-Reported and Objective Measures of Driving Exposure and Patterns in Older Drivers. *Accident Analysis & Prevention*, Vol. 42, 2010, pp. 523–529.
24. Huebner, K. D., M. M. Porter, and S. C. Marshall. Case Study Validation of an Electronic Device for Measuring Driving Exposure. *Traffic Injury Prevention*, Vol. 7, 2006, pp. 76–80.
25. Wood, J. M., P. F. Lacherez, and K. J. Anstey. Not All Older Adults have Insight into their Driving Abilities: Evidence from an An-Road Assessment and Implications for Policy. *Journals of Gerontology Series A: Biomedical Sciences and Medical Sciences*, Vol. 68, No. 5, 2013, pp. 559–566.
26. Silverstein, N. M., A. G. Gottlieb, D. W. Eby, L. J. Molnar, N. Kwan, T. Materdey, G. Adler, and E. Van Ranst. Do Memory-Impaired Drivers and Their Family Members Agree on Driving Ability and Fehaviors? *Transportation*

Research Record: Journal of the Transportation Research Board, 2011. 2265: 200–206.
27. Staplin, L., K. W. Gish, and J. Joyce. Low Mileage Bias and Related Policy Implications — A Cautionary Note. *Accident Analysis & Prevention*, Vol. 40, 2008, pp. 1249–1252.
28. Insurance Institute for Highway Safety. *Fatality Facts: General Statistics* [Internet], 2016. http://www.iihs.org/iihs/topics/t/general-statistics/fatalityfacts/overview-of-fatality-facts. Accessed July 29, 2017.
29. Eby, D. W., N. M. Silverstein, L. J. Molnar, D. LeBlanc, and G. Adler. Driving Behaviors in Early Stage Dementia: A Study Using In-Vehicle Technology. *Accident Analysis & Prevention*, Vol. 49, 2012, pp. 330–337.
30. Rimmö, P. A., and L. Hakamies-Blomqvist. Older Drivers' Aberrant Driving Behaviour, Impaired Activity, and Health as Reasons for Self-Imposed Driving Limitations. *Transportation Research Part F: Traffic Psychology and Behaviour*, Vol. 5, No. 1, 2002, pp. 345–360.
31. Heeringa, S. G., and J. H. Connor. *Technical Description of the Health and Retirement Survey Sample Design*. University of Michigan, Ann Arbor, MI, 1995.
32. Ofstedal, M. B., D. R. Weir, C. Kuang-Tsung, and J. Wagner. *Updates to HRS Sample Weights*. HRS Documentation Report. Survey Research Center, University of Michigan, Ann Arbor, MI, 2011.
33. Servais, M. *Overview of HRS Public Data Files for Cross-sectional and Longitudinal Analysis*. Institute for Social Research, University of Michigan, Ann Arbor, MI, 2004.
34. Chien, S., N. Campbell, C. Chan, O. Hayden, M. Hurd, R. Main, J. Mallett, C. McCullough, E. Meijer, M. Moldoff, and P. Pantoja. *RAND HRS Data Documentation, Version P*. Labor and Population Program, RAND Center for the Study of Aging, Santa Monica, CA, 2016.
35. Ofstedal, M. B., G. G. Fisher, and A. R. Herzog. *Documentation of Cognitive Functioning Measures in the Health and Retirement Study*. HRS/AHEAD Documentation Report. Institute for Social Research, University of Michigan, Ann Arbor, MI, 2005.
36. Health and Retirement Study. *Health and Retirement Study Codebook* [Internet]. Ann Arbor, MI, 2014. https://hrs.isr.umich.edu/documentation/codebooks. Accessed November 14, 2017.
37. Health and Retirement Study. *Sample Sizes and Response Rates* [Internet]. Ann Arbor, MI, 2011, pp. 1–13. http://hrsonline.isr.umich.edu/sitedocs/sampleresponse.pdf.
38. Lopez, O. L., and L. H. Kuller. Telephone Interview for Cognitive Status. *Neuroepidemiology*, Vol. 34, No. 1, 2010, pp. 63–64.
39. Brandt, J., M. Spencer, and M. Folstein. The Telephone Interview for Cognitive Status. *Neuropsychiatry Neuropsychol Behav Neurol*, Vol. 2, 1988, pp. 111–117.
40. Langa, K. M., E. B. Larson, E. M. Crimmins, J. D. Faul, D. A. Levine, M. U. Kabeto, and D. R. Weir. A Comparison of the Prevalence of Dementia in the United States in 2000 and 2012. *JAMA Internal Medicine*, Vol. 177, No. 1, 2017, pp. 51–58.
41. Crimmins, E. M., J. K. Kim, K. M. Langa, and D. R. Weir. Assessment of Cognition Using Surveys and Neuropsychological Assessment: The Health and Retirement Study and the Aging, Demographics, and Memory Study. *Journals of Gerontology Series B: Psychological Sciences and Social Sciences*, Vol. 66, 2011, i162–i171.
42. Freund B., and M. Szinovacz. Effects of cognition on driving involvement among the oldest old: variations by gender and alternative transportation opportunities. *Gerontologist*, Vol. 42, No. 5, 2002 Oct, pp. 621–33.
43. Choi, Moon, Matthew C Lohman, and Briana Mezuk. "Trajectories of Cognitive Decline by Driving Mobility: Evidence from the Health and Retirement Study." *Int J Geriatr Psychiatry* Vol. 29, No. 5, 2014, pp. 447–53.
44. Health and Retirement Study. *Cross-wave Geographic Information: Respondent Census Division/Degion and Mobility File 1992-2012, Version 5.1*. Health and Retirement Study, 2015.
45. Hanley, J. A. Statistical Analysis of Correlated Data Using Generalized Estimating Equations: An Orientation. *American Journal of Epidemiology*, Vol. 157, No. 4, 2003, pp. 364–375.
46. Zou, G. A Modified Poisson Regression Approach to Prospective Studies with Binary Data. *American Journal of Epidemiology*, Vol. 159, No. 7, 2004, pp. 702–706.
47. Barton, K. N. *The Impact of Social Network and Neighborhood Characteristics on the Driving Status of Older Adults: The Moderating Role of Gender*. Graduate doctoral dissertations. 2016, 281p.
48. Iliffe, S., L. Robinson, C. Brayne, C. Goodman, G. Rait, J. Manthorpe, and P. Ashley. Primary Care and Dementia: 1. Diagnosis, Screening and Disclosure. *Journal of Geriatric Psychiatry: A Journal of the Psychiatry of Late Life and Allied Sciences*, Vol. 24, 2009, pp. 895–901.
49. Nielsen, T. R., A. Vogel, T. K. T. Phung, A. Gade, and G. Waldemar. Over- and Under-Diagnosis of Dementia in Ethnic Minorities: A Nationwide Register-Based Study. *International Journal of Geriatric Psychiatry*, Vol. 26, 2011, pp. 1128–1135.
50. Tang, W., K. Kannaley, D. B. Friedman, V. J. Edwards, S. Wilcox, S. E. Levkoff, et al. Concern About Developing Alzheimer's Disease or Dementia and Intention to be Screened: An Analysis of National Survey Data. *Archives of Gerontology and Geriatrics*, Vol. 71, 2017, pp. 43–49.
51. Dow, J., M. Gaudet, and E. Turmel. Crash Rates of Quebec Drivers with Medical Conditions. *Annals of Advances in Automotive Medicine*, Vol. 57, 2013, pp. 57–66.
52. Meuleners, L. B., J. Ng, K. K. Chow, and M. Stevenson. Motor Vehicle Crashes and Dementia: A Population-Based Study. *Journal of the American Geriatrics Society*, Vol. 64, No. 5, 2016, pp. 1039–1045.
53. Fraade-Blanar, L. A., R. N. Hansen, K. C. G. Chan, J. M. Sears, H. J. Thompson, and P. K. Crane. Cognitive Decline and Older Driver Crash Risk. *Journal of the American Geriatrics Society*, Vol. 66, No. 6, 2018 pp. 1075–1981.
54. Pieper, B. Counseling the Patient Who is No Longer Safe to Drive. *Optometry*, 2006. pp. 93–96.
55. Choi, M., and B. Mezuk. Aging Without Driving: Evidence from the Health and Retirement Study, 1993 to 2008. *Journal of Applied Gerontology*, Vol. 32, No. 7, 2013, pp. 902–912.

56. Rosenbloom, S. Differences in Perceptions of Driving Skills Older Drivers and Adult Children of Older Drivers in the United Kingdom. *Transportation Research Record: Journal of the Transportation Research Board*, 2007. 2009: 15–22.
57. Connell, C. M., A. Harmon, M. R. Janevic, and L. P. Kostyniuk. Older Adults' Driving Reduction and Cessation: Perspectives of Adult Children. *Journal of Applied Gerontology*, Vol. 32, No. 8, 2012, pp. 975–996.
58. O'Connor, M. L., J. D. Edwards, V. G. Wadley, and M. Crowe. Changes in Mobility Among Older Adults with Psychometrically Defined Mild Cognitive Impairment. *Journals of Gerontology Series B: Psychological Sciences and Social Sciences*, Vol. 65, No. 3, 2010, pp. 306–316.
59. Knopman, D. S., R. O. Roberts, Y. E. Geda, V. S. Pankratz, T. J. H. Christianson, R. C. Petersen, and W. A. Rocca. Validation of the Telephone Interview for Cognitive Status-Modified in Subjects with Normal Cognition, Mild Cognitive Impairment, or Dementia. *Neuroepidemiology*, Vol. 34, No. 1, 2010, pp. 34–42.

The Standing Committee on Safe Mobility of Older Persons (ANB60) peer-reviewed this paper (18-03851).

Article

TRR
JOURNAL OF THE TRANSPORTATION RESEARCH BOARD

Transportation Research Record
2018, Vol. 2672(33) 101–108

DOI: 10.1177/0361198118798240
journals.sagepub.com/home/trr
SAGE

Newer Model Years Are Associated with Reduced Risk of Motor Vehicle Crash Fatalities among Older Drivers

Bernadette A. Fausto[1] and Brian C. Tefft[2]

Abstract

Older drivers are a growing and particularly vulnerable population. Advances in vehicle safety may reduce the risk of death in the event of a motor vehicle crash (MVC). Although previous studies have identified a relationship between increasing model years and decreased risk of death in MVC among adult drivers, this relationship has not been quantified among older drivers. The purpose of this study was to examine the relationship between vehicle model year and the risk of death among drivers aged 65 years and older who were involved in MVCs in the United States from 2010 to 2015. Data were from a census of all fatal MVCs and a representative sample of all police-reported crashes nationwide. Drivers were stratified into five age groups (65–69, 70–74, 75–79, 80–84, and 85+). Vehicles examined were cars, sport utility vehicles, pick-up trucks, and vans. Overall, older drivers driving vehicles older than model year 1990 were more than twice as likely to be killed in a crash as individuals driving vehicles of model years 2000 to 2009. Older drivers driving vehicles of model years 2010 and newer had 46% lower risk of MVC death than those driving vehicles of model year 2000 to 2009. The relationship between vehicle model year and risk of death in a crash varied by driver age and vehicle type, however, the same overall patterns were observed for all age groups and vehicle types examined. These findings suggest that older adults are responsive to improvements in vehicle safety.

The world's aging population will nearly double from 2012 to 2050 (*1*, *2*). Such changes in demography will lead to dramatic increases in the number of older drivers. In the U.S.A. alone, older drivers comprised 18% of all drivers in 2015, having increased from 15% in 2006 (*3*). The majority of persons aged 65 and older (86%) are current drivers (*4*) and by 2050 one in four drivers will be an older adult (*3*). Driving is a privilege typically earned in mid-adolescence and allows for participation in occupational and community activities in adulthood. Individuals remain reliant on automobile transportation even through retirement, rendering the ability to drive a lifelong activity.

Older adults are more prone to injury and death in motor vehicle crashes (MVCs) than younger age groups due to increased frailty (e.g., age-related reduction in fracture toughness of bone) (*5*). Given the growing numbers of older drivers and their increased risk of injury due to frailty (*6*, *7*), characterizing trends in the incidence of MVCs involving older adult merits continued investigation. Although MVC rates have decreased over the last decade (*8*), MVCs remain a leading cause of death among older adults. Between 2010 and 2015, in the U.S.A., MVCs were the second leading cause of unintentional injury deaths for individuals over 65, after unintentional falls (*9*, *10*).

Fortunately, advances in vehicle safety design, including crashworthiness and crash avoidance features, have prevented injuries and saved lives nationwide since the late 1960s (*11*). New and emerging technologies implemented in the last 20 years have the potential to further enhance public safety (*12*) as evidenced by the finding that with more recent model years the likelihood of being injured or fatally injured decreases (*13–16*). Blows et al. (*13*) reported that occupants of vehicle models older than 1984 had three times the risk of MVC injury compared with occupants of newer vehicle models. More recently, the National Highway Traffic Safety Administration found that the likelihood of escaping a MVC uninjured increased from 79% in 2000 to 82% in 2008 (*14*). Another study found that an increase in model year (i.e., newer models) was associated with decreased odds of

[1]School of Aging Studies, University of South Florida, Tampa, FL
[2]AAA Foundation for Traffic Safety, Washington, D.C.

Corresponding Author:
Address correspondence to Bernadette A. Fausto: bfausto@usf.edu

thoracic injuries, spinal injuries, and fatalities (*15*). Ryb and colleagues also reported that the odds of death decreased with later model years (*16*). Interestingly, one study found that more recent model years were associated with an increased risk of fatalities (*17*); however, the literature overwhelmingly supports the generalization that newer model years have a protective effect in relation to injuries and death.

Despite the mounting evidence of prevention of fatal MVCs due to advancements in car safety, most studies have examined only injuries and fatalities in broad adult age cohorts (e.g., 25–64, 65+) with little attention to these outcomes within more narrow older adult age groupings (*18*). Age is an independent predictor of MVC injury and fatality due to increased frailty (*2*, *19*, *20*); however, the association between age and physical health becomes increasingly variable, especially with advancing ages (*2*). Thus, it is important to explore the effect of newer models on MVC injury and fatality within more narrow age groupings since advancements in car safety and emerging safety technologies may reflect differences in MVC outcomes across age groups.

Of particular interest for the current study is the association of the model year with the odds of MVC fatality among older drivers. Compared with younger adult drivers, the rate of MVC deaths has decreased more slowly in older adults, which suggests that older adults may be less responsive to changes in car safety technology (*21*). The unique needs, heterogeneity in biophysical factors, and the impact of crash protection technologies on older drivers warrants investigation.

Therefore, the purpose of the current study was to examine the association of model year with MVC fatality risk of older drivers aged 65 and over, to test the hypothesis that newer models would confer a protective effect on MVC fatalities among older drivers involved in crashes. Such findings will inform future safety technology development and road safety policy strategies geared towards older adults.

Method

Data

Data were obtained from the National Automotive Sampling System/General Estimates System (GES) and Fatal Accident Reporting System (FARS) databases. Both databases were developed by the National Highway Traffic Safety Administration (NHTSA) with the goals of providing measures on highway safety, identifying road safety problems and patterns, and aiding in the development of safety solutions.

The GES database contains comprehensive data on police-reported MVCs ranging from minor property damage with no significant personal injury to fatalities. These data have been collected annually since 1988. Selected variables include time, location, injury severity, seatbelt use, drug or alcohol involvement, number of vehicles involved, number of people involved as well as their demographics, and the role of people involved. Each year, the GES randomly selects approximately 54,000 crashes from a stratified sample of jurisdictions nationwide. Weights are assigned to each crash to derive national estimates. Full details of the GES methodology can be accessed on the NHTSA website (https://www.nhtsa.gov/research-data/national-automotive-sampling-system-nass).

FARS data contains information on all MVC fatalities in the U.S.A. since 1975. Each of the 50 states, the District of Columbia, and Puerto Rico send official state records to NHTSA on crash aspects similar to those listed for GES. To be included in FARS, the crash must involve a fatality that occurred within 30 days after the crash. Comprehensive details on the FARS database are available on the NHTSA website (https://www.nhtsa.gov/research-data/national-automotive-sampling-system-nass).

Procedure

The current analysis used both GES and FARS aggregate data for the years 2010–2015. The analysis focused only on drivers to control for the effect of seating position (*17*). For both GES and FARS databases, drivers 64 and younger and drivers with no reported age were excluded. The included drivers were categorized into the following age groups: 65–69, 70–74, 75–79, 80–84, and 85+.

The model years for cars were grouped as follows : <1990, 1990–1999; 2000–2009, and ⩾2010. Vehicle types examined were automobiles and auto derivatives, sport utility vehicles (SUVs), van-based light trucks, and pick-up trucks. Van-based light trucks included vehicle types such as large vans, walk-in vans, and light-truck-based motorhomes and are referred to hereafter as vans. Other types of vehicles (e.g., motorcycles, golf carts, large trucks, buses) were excluded.

The response variable in this study was MVC driver death (coded as 4 – Fatal Injury from FARS), and all other codes indicating injury (1 – Possible Injury, 2 – Suspected Minor Injury, 3 – Suspected Serious Injury, and 5 – Injured, Severity Unknown) or no injury (0 – No Apparent Injury and 6 – Died Prior to Crash) were coded as not killed. Thus, all driver deaths were coded as 0 (not killed) or 1 (killed). Drivers involved in fatal crashes reported in the GES database were excluded to avoid double counting as all fatal crashes are also included in FARS.

Statistical Analysis

Descriptive cross-tabulations were conducted for select demographic (age, sex) and vehicle (model year, vehicle type) characteristics for drivers involved in non-fatal crashes (from GES) and drivers involved in fatal crashes (from FARS) separately.

Multiple Poisson regression analysis was used to investigate the relationship between vehicle model year and the relative risk of death in a crash while controlling for the effects of driver age, vehicle type, calendar year, and interactions of vehicle model year with driver age group and vehicle type. The statistical significance of interactions was evaluated using Wald *F*-tests. Model year group 2000–2009 was treated as the reference group. From the Poisson regression analysis, the relative risk (RR) and 95% confidence interval were calculated. Statistical significance was concluded when the 95% confidence interval did not overlap with RR = 1 ($p < .05$). Variances were estimated using the robust variance estimator (*22*).

Descriptives and statistical analysis were performed using Stata software, version 15.0 (StataCorp, LLC, College Station, TX). All statistics reported on drivers involved in non-fatal crashes are based on weighted GES data unless noted otherwise. Estimation of standard errors, confidence intervals, and *p*-values account for the stratified sample design of the GES data.

Results

Descriptive Statistics

Table 1 shows the number of eligible older drivers for GES and FARS 2010–2015 study files after excluding individuals younger than 65, non-drivers, and drivers of excluded car vehicle types (e.g., buses, motorcycles). The numbers of eligible GES older drivers per age group are presented as unweighted and weighted estimates.

Demographic characteristics of eligible older drivers and their vehicles from GES and FARS are presented in Tables 2 and 3, respectively.

Table 1. Number of Eligible Older Drivers in GES and FARS Study Files (2010–2015)

	GES		
Age group	Unweighted	Weighted	FARS
65–69 years	15,638	1,875,982	14,554
70–74 years	10,556	1,228,237	11,757
75–79 years	7,363	866,663	10,039
80–84 years	5,201	558,387	8,443
85+ years	3,796	423,432	7,074
Total	42,554	4,982,701	51,867

For GES, the mean age of older drivers was 73.38. As expected, the majority of older drivers across all older age groups were male (comprising 56.17% of the GES sample). Model year group 2000–2009 represented the majority of the vehicles from MVCs between 2010 and 2015. The majority of older drivers were driving automobiles and auto-derivatives (62.26%).

For FARS, the mean age of older drivers was 75.22. The majority of drivers were male across all age groups (66.60%). Over half of the vehicles (56.52%) were within the model year group 2000–2009. Just 3.71% of the vehicles involved in MVCs during the years examined were in model year group <1990. Similar to GES, the majority of older drivers were driving automobiles and auto-derivatives (56.59%).

Multiple Poisson Regression Results

Model Year and Age Group Analysis. Multiple Poisson regression analysis was used to investigate the relationship between vehicle model year and the RR of death in a crash while controlling for the effects of driver age, vehicle type, calendar year, and interactions of vehicle model year with driver age group and vehicle type.

The effects of model year and age group on the likelihood of dying in a MVC are quantified and summarized in the first five rows of Table 4. The relationship between model year and risk of driver death varied by age group ($p < 0.001$; Wald test of interaction of model year × age group). Overall, drivers in vehicles of model years older than 2000 had increased risk of MVC death compared with drivers in vehicles of model years 2000–2009. Drivers in vehicles of model years 2010 and newer had reduced risk of MVC deaths compared with drivers of model year 2000–2009 vehicles. The RR of 2.72, which is that displayed for the group aged 65–69, indicates a driver aged 65–69 had 2.72 times the risk of death if involved in a crash in a vehicle of model year 1990 or older compared with a vehicle of model year 2000–2009 ($p < .001$). The risk of MVC death of a 65–69 year-old driver in a vehicle of model year 1990–1999 was 1.83 times that of a driver in a vehicle of model year 2000–2009 ($p < .001$). For the same age group, the RR of 0.53 reflects that the risk of death is 47% lower if involved in a crash in a vehicle of model year 2010 or newer compared with model years 2000–2009 ($p < .001$).

For drivers aged 75–79, the risk of death for model years older than 1990 was 2.56 times as likely than for model years 2000–2009 ($p < .001$). The risk of death for model years 1990–1999 was 1.49 times that of model years 2000–2009 ($p < .001$). Within the same age group, model years 2010 and newer demonstrated 47% less risk of MVC driver death compared with model years 2000–2009 ($p < .001$).

Table 2. Demographic Characteristics of Older Drivers and Their Vehicles in Non-Fatal MVCs by Age Category (GES, 2010–2015)

Characteristics	65–69	70–74	75–79	80–84	85+
Female (%)	44.07	43.82	43.10	44.33	43.61
Model year					
<1990	1.92	1.88	1.79	1.32	2.13
1990–1999	15.44	15.97	18.31	18.02	22.79
2000–2009	57.79	58.51	58.43	62.44	57.99
>2010	24.85	23.65	21.48	18.23	17.09
Vehicle type					
Auto/auto-derivatives	56.58	58.95	65.19	70.67	79.29
SUV	20.19	18.37	14.93	11.90	8.15
Van-based light truck	7.02	7.99	7.44	7.66	4.80
Pick-up truck	16.21	14.70	12.44	9.77	7.76

Note: Column percentages are displayed for each age category for given characteristic. SUV = sport utility vehicle.

Table 3. Demographic Characteristics of Older Drivers and their Vehicles in Fatal MVCs by Age Category (FARS, 2010–2015)

Characteristics	65–69	70–74	75–79	80–84	85+
Female (%)	32.61	32.36	34.41	34.40	34.33
Model year					
<1990	3.56	3.48	3.64	3.68	4.48
1990–1999	22.30	24.08	23.53	25.93	28.91
2000–2009	56.21	56.09	58.30	56.58	55.44
>2010	17.93	16.35	14.52	13.81	11.16
Vehicle type					
Auto/auto-derivatives	46.76	50.89	57.87	65.40	74.84
SUV	19.78	17.39	14.63	10.59	8.20
Van-based light truck	8.99	8.79	8.65	7.78	5.24
Pick-up truck	24.47	22.94	18.85	16.23	11.71

Note: Column percentages are displayed for each age category for given characteristic. SUV = sport utility vehicle.

In the oldest age group (85+), the risk of death for vehicle model years older than 1990 was 2.51 times as likely than for model years 2000–2009 ($p < .001$). The risk of death for model years 1990–1999 was 1.42 times that of model years 2000–2009 ($p < .001$). A driver in a vehicle of model year 2010 or newer was at 37% reduced risk of MVC death.

Model Year and Vehicle Type Analysis. Similarly, the relationship between model year and risk of driver death varied by vehicle type ($p < 0.001$; Wald test of interaction of model year × vehicle type). Thus, the last four rows in Table 4 display the RRs and 95% CIs for each vehicle type group (automobiles and auto derivatives, SUVs, vans, and pick-up trucks). Overall, across all vehicle types, model years older than 2000 were associated with increased risk of MVC death whereas model years 2010 and newer were associated with reduced risk of MVC death.

For automobiles and auto-derivatives of model years older than 1990, the risk of MVC death was 3.26 times higher than for model years 2000–2009 ($p < .001$). The risk of death for model years older than 1990 was 1.75 times that of model years 2000–2009 ($p < .001$). For the same vehicle type, the RR of 0.58 reflects that the risk of driver death is 42% lower if involved in a crash in a vehicle of model year 2010 compared with model years 2000–2009 ($p < .001$).

SUVs of model years older than 1990 were associated with 2.87 times the risk of MVC driver death compared with model years 2000–2009 ($p < .001$). The risk of death for model years 1990–1999 was 2.05 times that of model years 2000–2009 ($p < .001$). The risk of death was reduced by more than half (51%) for model years 2010 and newer compared with model years 2000–2009 ($p < .001$).

The risk of driver death in MVCs for vans older than 1990 was 1.66 times that of model years 2000–2009. Similarly, drivers in vans of model years 1990–1999 had

Table 4. Risk Ratios and 95% Confidence Intervals for Driver Death in Relation to Vehicle Model Year

	Model year (reference = Model year 2000–2009)		
Characteristic	<1990	1990–1999	>2010
Overall	2.70 (2.34–3.11)	1.77 (1.64–1.91)	0.54 (0.50–0.58)
Age group[a]			
65–69 years	2.72 (2.24–3.30)	1.83 (1.66–2.01)	0.53 (0.49–0.57)
70–74 years	2.52 (1.97–3.24)	1.98 (1.77–2.20)	0.53 (0.48–0.58)
75–79 years	2.56 (1.99–3.29)	1.49 (1.34–1.65)	0.53 (0.47–0.60)
80–84	3.91 (2.84–5.40)	1.85 (1.65–2.08)	0.70 (0.62–0.78)
85+ years	2.51 (2.02–3.11)	1.42 (1.26–1.60)	0.63 (0.55–0.72)
Body type[a]			
Auto/auto derivative	3.26 (2.82–3.76)	1.75 (1.62–1.89)	0.58 (0.54–0.63)
Sport utility vehicle	2.87 (1.71–4.83)	2.05 (1.78–2.37)	0.49 (0.44–0.53)
Van-based light truck	1.66 (1.18–2.34)	1.66 (1.46–1.90)	0.54 (0.46–0.63)
Pick-up truck	1.72 (1.25–2.39)	1.53 (1.34–1.73)	0.51 (0.44–0.59)

[a]Poisson regression models adjusted for calendar year.

1.66 times the risk of death in MVCs compared with those in model years 1990–1999 ($p < .001$). The risk of MVC driver death was reduced by 46% for models 2010 and newer compared with model years 2000–2009 ($p < .001$).

Finally, for pick-up trucks older than 1990, the risk of MVC death was 1.72 times that of model years 2000–2009 of the same vehicle type ($p < .001$). The risk of MVC driver death for pick-up trucks of model years 1990–1999 was 1.53 times that of model years 2000–2009 ($p < .001$). Pick-up trucks 2010 and newer were associated with 49% reduced risk of driver death in MVCs ($p < .001$).

Discussion

The findings in this paper demonstrate that the risk of MVC death decreased with newer model years and that this varied by age group and vehicle type. Overall, across all age groups, model years older than 2000 were associated with increased risk of MVC driver death, whereas model years 2010 and newer conferred reduced risk of driver death. Similarly, across the four vehicle type categories, model years older than 2000 and model years 2010 and newer showed increased and reduced risk of MVC driver death, respectively. Older adults appear to be responding to improvements in vehicle safety. Thus, the hypothesis that newer models would confer a protective effect on MVC death among older drivers was supported.

These findings are consistent with previous literature that newer model years are associated with decreased odds of MVC death in adult front seat occupants 18 years of age or older and front seat occupants aged 16 years or older (*15*, *16*). Ryb and colleagues stated that the observed decrease in MVC deaths over the past few decades can be attributed not only to the attenuated number and severity of crashes but also the mitigation of harm experienced by vehicle occupants (*16*). The effect of harm reduction perhaps attributed to improved vehicle safety appears to benefit drivers or front seat occupants, regardless of age.

The findings also show that the relationship between model year and MVC death varied by vehicle type. Across all vehicle types, however, newer models were associated with reduced odds of MVC death and older models were associated with increased odds of MVC death. Odds of death were reduced by nearly half for models older than 2010 for all vehicle types.

The protective effect of newer model-year vehicles may be attributed to improvements in occupant protection/vehicle safety including crash avoidance, crashworthiness, and postcrash assistance (*23*). One approach to prevention of MVC deaths includes standard car safety features such as side airbags and electronic stability control (ESC). Side-impact collisions accounted for nearly a third of all MVC deaths in 2004 (*24*). Encouragingly, side airbags have been found to significantly reduce driver death risk for males and females, older and younger drivers, and drivers of small and midsize cars (*25*). The majority of vehicle models from 2008 and later have standard side airbags. A more recent standard safety feature is ESC, which prevents the vehicle from sliding or skidding while turning or negotiating a curve. ESC was found to reduce the risk of MVC death by 23% (*26*) and estimated to save roughly 1,500 to 2,500 lives and prevent between 46,000 and 65,800 injuries per year in the U.S.A. (*27*). A literature review revealed strong evidence for the effectiveness of ESC in reducing fatal MVCs up to 50% in cars and up to 70% in SUVs (*28*). In 2012, the NHTSA mandated all passenger vehicles, trucks, and buses under 10,000 pounds to be equipped with ESC (*29*).

Interestingly, while the RRs across the five age groups for model years 2010 and newer reflected reduced risk of driver death in MVCs, drivers in the oldest age groups (80–84 and 85+) appear to gain less benefit from improvements in vehicle safety compared with the other age groups. The risk of death across the 65–69, 70–74, and 75–79 year-old drivers involved in MVCs in vehicles of model years 2010 and newer was almost half that for model years 2000–2009. For the oldest drivers, however, those aged 80–84 and 85+, the risk of death from MVCs in vehicles of model years 2010 and newer was reduced by 30% and 37%, respectively, compared with model years 2000–2009. Thus, future research should examine factors related to the comparatively lower responsiveness to improvements in vehicle safety among drivers aged 80 or older.

This study is not without limitations. While the relationship between vehicle model year and the relative risk of death in a crash while controlling for the effects of driver age, vehicle type, and calendar year was quantified, other factors may have contributed to this relationship. External factors such as healthcare accessibility, timing of emergency services, and improvements in medical care are potential confounders. Adjusting for calendar year may have captured these effects. However, if older versus newer vehicles tend to be driven under different traffic or environmental characteristics that might also influence the risk of death in the event of a crash, such factors might confound the relationship between vehicle model year and risk of death. Additionally, similar to previous studies, model year appears to be used as a proxy for improvements in vehicle safety (*13–16*), but does not provide information about the mechanism by which newer vehicles reduce risk of death for older drivers. The availability of vehicle safety technologies (e.g., ESC, lane departure warnings) and the vehicle crashworthiness ratings (e.g., from NHTSA or the Insurance Institute for Highway Safety) are currently outside the scope of GES and FARS data collection and beyond the scope of the current study. Future research could attempt to examine the specific mechanisms by which newer vehicles reduce the risk of death of older drivers in the event of a crash.

Although uncovering the specific mechanisms by which newer model year vehicles have reduced the risk of death for older drivers involved in crashes was beyond the scope of this study, it is possible that newer vehicles better protect their occupants from MVC fatality, given involvement in the same crash, due to technologies such as side air bags as well as improvements in the occupant compartment's resistance to intrusion in the event of a crash. Indeed, side airbags appear to disproportionately benefit older drivers (versus younger drivers), a group that is typically overrepresented in side-impact crashes, especially at intersections (*30*). This possible mechanism could be investigated by appending vehicle crash test ratings (e.g., from NHTSA or the Insurance Institute for Highway Safety) to the data analyzed for the current study and examining whether vehicle model year remained significantly associated with fatality risk after accounting for safety ratings.

It is also possible that active safety technologies, such as ESC, have changed the distribution of crash severity by disproportionately reducing the rates of types of crashes that would have been of greater severity had they occurred (e.g., high-speed road departure crashes). This mechanism seems less likely, however, as this study focused on older drivers, and older drivers are rarely involved in high-speed road departure crashes. This could be examined in more granular analyses in which data were stratified by crash type (e.g., crash geometry, single- versus multi-vehicle crash).

Conclusions

Older drivers will become increasingly prevalent on the road as the baby boomer population ages. This trend is particularly alarming, because older drivers are a vulnerable population in MVCs compared with younger drivers due to their increased frailty and poorer physical health. However, within the population of older drivers alone, physical health becomes increasingly variable. Previous studies have not attempted to examine the effect of model year on MVC death among older drivers. The current study found that, in the event of a crash, older drivers have much lower risk of death if they are driving newer model vehicles.

Acknowledgments

This work was supported by the AAA Foundation for Traffic Safety. The authors acknowledge Tara Kelley-Baker, Woon Kim, Lindsay Arnold, and Pedro Adorno for their helpful feedback on the manuscript.

Author Contributions

The authors confirm contribution to the paper as follows: study conception and design: Brian C. Tefft, Bernadette A. Fausto; data extraction: Brian C. Tefft, Bernadette A. Fausto; analysis and interpretation of results: Brian C. Tefft, Bernadette A. Fausto; draft manuscript preparation: Brian C. Tefft, Bernadette A. Fausto. All authors reviewed the results and approved the final version of the manuscript.

References

1. Ortman, J. M., and V. A. Velkoff. *An Aging Nation: The Older Population in the United States*. Washington, D.C., 2014.
2. Newgard, C. D. Defining the "Older" Crash Victim: The Relationship Between Age and Serious Injury in Motor Vehicle Crashes. *Accident Analysis and Prevention*, Vol. 40, No. 4, 2008, pp. 1498–1505.
3. National Highway Traffic Safety Administration. *Traffic Safety Facts: 2015 Data*. Publication DOT HS 812 372, National Highway Traffic Safety Administration, Washington, D.C., 2015.
4. Rosenbloom, S., and R. Santos. *Understanding Older Drivers: An Examination of Medical Conditions, Medication Use, and Travel Behavior*. AAA Foundation for Traffic Safety, Washington, D.C., 2014.
5. Wang, X., X. Shen, X. Li, and C. M. Agrawal. Age-related Changes in the Collagen Network and Toughness of Bone. *Bone*, Vol. 31, No. 1, 2002, pp. 1–7.
6. Anstey, K. J., J. Wood, S. Lord, and J. G. Walker. Cognitive, Sensory and Physical Factors Enabling Driving Safety in Older Adults. *Clinical Psychology Review*, Vol. 25, No. 1, 2005, pp. 45–65.
7. McGwin, G., R. V. Sims, L. Pulley, and J. M. Roseman. Relations Among Chronic Medical Conditions, Medications, and Automobile Crashes in the Elderly: A Population-based Case-control Study. *American Journal of Epidemiology*, Vol. 152, No. 5, 2000, pp. 424–431.
8. Kramarow, E., L. Chen, H. Hedegaard, and M. Warner. *Deaths from Unintentional Injury Among Adults Aged 65 and Over: United States, 2000–2013*. National Center for Health Statistics, Hyattsville, MD, 2015.
9. National Center for Injury Prevention and Control. *10 Leading Causes of Injury Deaths by Age Group Highlighting Unintentional Injury Deaths, United States – 2010*. National Center for Injury Prevention and Control, Atlanta, GA, 2010.
10. National Center for Injury Prevention and Control. *10 Leading Causes of Injury Death by Age Group Highlighting Unintentional Injury Deaths, United States*. Centers for Disease Control and Prevention, Atlanta, GA, 2015.
11. O'Neill, B. Preventing Passenger Vehicle Occupant Injuries by Vehicle Design – A Historical Perspective. *Traffic Injury Prevention*, Vol. 10, 2009, pp. 113–126.
12. Glassbrenner, D. *An Analysis of Recent Improvements to Vehicle Safety*. National Highway Traffic Safety Administration, Washington, D.C., 2012.
13. Blows, S., R. Q. Ivers, M. Woodward, J. Connor, S. Ameratunga, and R. Norton. Vehicle Year and the Risk of Car Crash Injury. *Injury Prevention*, Vol. 9, No. 4, 2009, pp. 353–356.
14. National Highway Traffic Safety Administration. How Vehicle Age and Model Year Relate to Driver Injury Severity in Fatal Crashes. US Department of Transportation, Washington, D.C., 2013.
15. Ryb, G. E., P. C. Dischinger, and S. Ho. Vehicle Model Year and Crash Outcomes: A CIREN Study. *Traffic Injury Prevention*, Vol. 10, No. 3, 2009, pp. 560–566.
16. Ryb, G. E., P. C. Dischinger, G. McGwin, and R. L. Griffin. Crash-related Mortality and Model Year: Are Newer Vehicles Safer? *Annals of Advances in Automotive Medicine*, Vol. 55, 2011, pp. 113–121.
17. Bédard, M., G. H. Guyatt, M. J. Stones, and J. P. Hirdes. The Independent Contribution of Driver, Crash, and Vehicle Characteristics to Driver Fatalities. *Accident Analysis and Prevention*, Vol. 34, No. 6, 2002, pp. 717–727.
18. Yoganandan, N., F. A. Pintar, B. D. Stemper, T. A. Gennarelli, and J. A. Weigelt. Biomechanics of Side Impact: Injury Criteria, Aging Occupants and Airbag Technology. *Journal of Biomechanics*, Vol. 40, No. 2, 2007, pp. 227–243.
19. O'Donnell, C. J., and D. H. Connor. Predicting the Severity of Motor Vehicle Accident Injuries Using Models of Ordered Multiple Choice. *Accident Analysis and Prevention*, Vol. 28, No. 6, 1996, pp. 739–753.
20. Kononen, D. W., C. A. C. Flannagan, and S. C. Wang. Identification and Validation of a Logistic Regression Model for Predicting Serious Injuries Associated with Motor Vehicle Crashes. *Accident Analysis and Prevention*, Vol. 43, No. 1, 2011, pp. 112–122.
21. Lotfipour, S., R. Sayegh, B. Chakravarthy, W. Hoonpongsimanont, C. L. Anderson, J. C. Fox, and F. E. Vaca. Fatality and Injury Severity of Older Adult Motor Vehicle Collisions in Orange County, California, 1998–2007. *Western Journal of Emergency Medicine*, Vol. 14, No. 1, 2013, pp. 63–68.
22. Hardin, J. W., and J. M. Hilbe. *Generalized Linear Models and Extensions*, 2nd ed. Stata Press, College Station, TX, 2007.
23. Pike, J. Reducing Injuries and Fatalities to Older Drivers: Vehicle Concepts. In *Transportation in an Aging Society: A Decade of Experience*. Transportation Research Board, Bethesda, MD, 2004, pp. 213–226.
24. National Highway Traffic Safety Administration. *Traffic Safety Facts 2004: A Compilation of Motor Vehicle Crash Data from the Fatality Analysis Reporting System and the General Estimates System*. National Center for Statistics and Analysis, US Department of Transportation, Washington, D.C., 2004.
25. McCartt, A. T., and S. Y. Kyrychenko. Efficacy of Side Airbags in Reducing Driver Deaths in Driver-side Car and

SUV Collisions. *Traffic Injury Prevention*, Vol. 8, No. 2, 2007, pp. 162–170.

26. Sivinski, R. *Crash Prevention Effectiveness of Light-vehicle Electronic Stability Control: An Update of the 2007 NHTSA Evaluation*. National Highway Traffic Safety Administration, Washington, D.C., 2011.
27. National Highway Traffic Safety Administration. *Electronic Stability Control Systems*. National Highway Traffic Safety Administration, Washington, D.C., 2007.
28. Ferguson, S. A. The Effectiveness of Electronic Stability Control in Reducing Real-world Crashes: A Literature Review. *Traffic Injury Prevention*, Vol. 8, No. 4, 2007, pp. 329–338.
29. National Highway Traffic Safety Administration. *Federal Motor Vehicle Safety Standards; Electronic Stability Control Systems*. Controls and Displays. No. 2127AJ77. National Highway Traffic Safety Administration, Washington, D.C., 2012.
30. Kahane, C. J. *Injury Vulnerability and Effectiveness of Occupant Protection Technologies for Older Occupants and Women*. Report No. DOT HS 811 766. National Highway Traffic Safety Administration, Washington, D.C., 2013.

The Standing Committee on Safe Mobility of Older Persons (ANB60) peer-reviewed this paper (18-05589).

Sage Publications USA
sagepub.com

CURRAN ASSOCIATES INC.
proceedings.com

ISBN 978-1-5108-7835-8

9 781510 878358